Pheno-phospholipids and Lipo-phenolics

Mohamed Fawzy Ramadan

Pheno-phospholipids and Lipo-phenolics

Novel Structured Antioxidants

 Springer

Mohamed Fawzy Ramadan
Agricultural Biochemistry Department
Zagazig University, Faculty of Agriculture
Zagazig, Egypt

ISBN 978-3-030-67401-4   ISBN 978-3-030-67399-4   (eBook)
https://doi.org/10.1007/978-3-030-67399-4

This Springer imprint is published by the registered company Springer Nature Switzerland AG
The registered company address is: Gewerbestrasse 11, 6330 Cham, Switzerland

Dedicated to my beloved family

Preface

Phospholipids (PL) and phenolic compounds are novel bioactive molecules with functional, nutritional, and health-promoting values. Phospholipids are commonly used as emulsifiers and antioxidants. Natural phenolics and flavonoids are novel bioactive compounds, but their biological properties are limited due to their hydrophilic character. To overcome this limitation, flavonoids and phytoextracts in combination with PL have been developed, which possess better absorption and functional traits. Phenolics-enriched PL (pheno-phospholipids) play an important role in enhancing the functional properties of both phenolic compounds and PL in food. Phenolics-enriched lecithin and PL-based drug delivery systems (phytosome) are promising as efficacious herbal drug delivery. In addition, the lipophilicity of phenolics could be modified by attaching a lipophilic moiety to the phenolic compound (lipo-phenolics) to alter its hydrophilic–lipophilic balance. Lipophilization allows the formation of new functionalized bioactive molecules having beneficial properties compared to natural hydrophilic phenolics. The lipophilization of phenolics enhances their solubility in apolar media. Lipophilization changes the location of the new compound in the emulsions, which might increase their antioxidant activities. Thus, in emulsified systems, lipo-phenolics are supposed to locate at the lipid/aqueous phase interface and to increase the protection of fats and oils. Lipo-phenolics proved to be excellent antioxidants in food and cosmetics.

Phenolipids have found several uses on an industrial scale, due to the costs, availability of raw material, and safety. Recent advances in the field of lipophilization allow accessing molecules with high potency and targeted action covering a wide spectrum of bioactivities. Phenolipids find applications in niche sectors (cosmetics and pharmaceutics) as well as in the novel food sector. Preparation, physicochemical characteristics, and functional properties of phenolipids and phytosomes are reviewed and discussed in this book. The latest developments and the current industrial status of phenolipids and phytosomes are described. Besides, this book reports on the chemistry, preparation, and functionality of lipid-enriched phenolics (lipo-phenolics), broadening their applications in food, pharmaceuticals, and cosmetics. The strategies of the lipophilization of phenolics, the effect of modification on the biological properties, and potential applications of the resulting lipo-phenolics are also reviewed.

This book was written upon kind invitation from Springer Nature. It contains chapters that describe chemistry, functionality, and techno-applications of pheno-phospholipids and lipo-phenolics. With the goal to provide a major reference work for those involved with pharmaceuticals, nutraceuticals, and the edible oil industry as well as undergraduate and graduate students, this volume presents a comprehensive review of the results that have led to the advancements in phenolipid chemistry, technology, and applications. I hope that the book will be a valuable source for people involved in pharmaceuticals, nutraceuticals, medical plants, and functional food.

The help and support given to me by the Springer Nature staff, especially *Daniel Falatko* and *Arjun Narayanan*, was essential for the completion of my task and is appreciated.

Let food be your medicine and medicine be your food. (Hippocrates)

Makkah, Saudi Arabia   Mohamed Fawzy Ramadan
October 2020

About the Book

Phenolic compounds are powerful bioactive compounds, but their use as antioxidant agents is limited due to their hydrophilic trait. The poor solubility of phenolics in apolar media limits their bioavailability and functional applications. A promising technique to overcome low solubility of phenolics is to increase their hydrophobicity by grafting with lipophilic moieties to formulate lipid-enriched phenolics (lipophenolics). Another way to enhance the amphiphilic traits of phenolics is by lipophilization with phospholipids (PL) to form phenolics-enriched phospholipids (pheno-phospholipids). Both functionalized molecules (phenolipids) exhibit unique bioavailability and functional properties.

Phenolipids have found several applications, probably due to the cost, availability of starting material, and safety. Recent advances in the field of lipophilization allow accessing molecules with high potency and targeted action covering a wide spectrum of bioactivities. Owing to their cost and availability, phenolipids find applications in niche sectors (cosmetics and pharmaceutics) as well as in the novel food sector.

Pheno-phospholipids and Lipo-phenolics: Novel Structured Antioxidants cover several specific topics with a focus on the strategies in the lipophilization of phenolics, the effect of lipophilization on the functional properties, and potential applications of functionalized phenolipids. The book brings a diversity of developments in food science to chemists, nutritionists, and students in food science, nutrition, lipids chemistry and technology, pharmaceuticals, cosmetics, and nutraceuticals.

Pheno-phospholipids and Lipo-phenolics: Novel Structured Antioxidants is a key textbook for pharmaceutical and functional food developers as well as research and development (R&D) managers working in all sector using medical plants, natural antioxidants, and edible oils. It is a useful reference work for companies reformulating their products or developing new products.

Key Features

- Specific information on phenolipids as novel structured antioxidants
- Preparation, physicochemical characteristics, and functional properties of phenolipids
- Latest developments and industrial applications of phenolipids
- Report on the chemistry and functionality of lipid-enriched phenolics (lipo-phenolics)
- Report on the chemistry and functionality of phenolics-enriched phospholipids (pheno-phospholipids)
- Strategies for the lipophilization of phenolic compounds
- Effect of lipophilization on the biological properties, and applications of lipo-phenolics

Readership

- Academics and students with a research interest in the area (pharmacologists, food chemists, lipid scientists, and food scientists).
- Pharmaceutics, functional food developers, and R&D managers working in all sectors using functional food, nutraceuticals, pharmaceuticals, and specialty oils.

Contents

About the Author

Mohamed Fawzy Ramadan is a Professor of Biochemistry in the Agricultural Biochemistry Department at Zagazig University, Egypt. Since 2013, he is a Professor of Biochemistry and consultant of international publishing at the Deanship of Scientific Research (Umm Al-Qura University, Makkah, Saudi Arabia).

In 2004, Prof. Ramadan obtained his Ph.D. (*Dr.rer. nat.*) in food chemistry from Berlin University of Technology (Germany). Prof. Ramadan continued his postdoctoral research in ranked universities in different countries, such the University of Helsinki (Finland), Max-Rubner Institute (Germany), Berlin University of Technology (Germany), and University of Maryland (USA). In 2010, he was appointed as visiting professor (100% research) at King Saud University (Saudi Arabia). In 2012, he was appointed as visiting professor (100% teaching) in the School of Biomedicine at Far Eastern Federal University in Vladivostok (Russian Federation).

Prof. Ramadan has published more than 250 research papers and conducts reviews for international peer-reviewed journals. He has edited and published several books and book chapters (Scopus h-index is 40 and more than 4500 citations). He was an invited speaker at several international conferences. Since 2003, Prof. Ramadan is a reviewer and editor in several highly cited international journals such as the *Journal of Medicinal Food* and *Journal of Advanced Research*.

Prof. Ramadan has received Abdul Hamid Shoman Prize for Young Arab Researcher in Agricultural Sciences (2006); Egyptian State Prize for Encouragement in Agricultural Sciences (2009); European Young Lipid Scientist Award (2009); AU-TWAS Young Scientist National Awards (Egypt) in Basic Sciences, Technology and Innovation (2012); TWAS-ARO Young Arab Scientist (YAS) Prize in Scientific and Technological Achievement (2013); and Atta-ur-Rahman Prize in Chemistry (2014).

Chapter 1
Chemistry and Functionality of Phenolipids

Abstract Natural phenolics are potent bioactive compounds, but their use as antioxidant agents in lipid-based foodstuffs and cosmetics is limited due to their hydrophilic trait. The low solubility of phenolics in apolar media limits their bioavailability and functional applications. In emulsion, phenolics are located in the aqueous compartment, while the lipid oxidation reactions occur at the oil-water interface. A promising technique to overcome phenolics' low solubility is to increase phenolics' hydrophobicity by grafting with lipophilic moieties to form lipid-enriched phenolics (lipo-phenolics). Another method to enhance the amphiphilic traits of phenolic compounds by lipophilization with phospholipids in a suitable solvent to form phenolics-enriched phospholipids (pheno-phospholipids). The resulting functionalized phenolics (phenolipids) exhibits higher bioavailability and antioxidative potential. This book will report on the strategies used in the lipophilization of phenolics, the impact of lipophilization on the functional traits, and fruitful applications of functionalized phenolipids.

Keywords Pheno-phospholipids · Lipophilization · Phenolics · Lipo-phenolics, functional properties · Antioxidants

Introduction

Phenolics are natural secondary metabolites with several biological effects contributing to the host defense system. Most of the natural antioxidants from plant sources are polar molecules, including phenolic acids, flavonoids, vitamin C, and diterpenes. The interest in phenolics has risen because of their functional traits such as antiallergic, antioxidant, anti-inflammatory, antimicrobial, antiviral, and anticarcinogenic properties. Synthetic antioxidants such as butylated hydroxytoluene (BHT), tert-butyl hydroquinone (TBHQ), and butylated hydroxyanisole (BHA) are used to protect food from rancidity. However, synthetic antioxidants are linked with harmful health effects and the incidence of carcinogenesis. The health risks have restricted the use of synthetic antioxidants in diet, wherein a high interest in plant extracts and natural phenolics are considered (Szydłowska-Czerniak, Rabiej, & Krzemiński, 2018;

M. F. Ramadan, *Pheno-phospholipids and Lipo-phenolics*,
https://doi.org/10.1007/978-3-030-67399-4_1

Szydłowska-Czerniak, Rabiej, Kyselka, Dragoun, & Filip, 2018). The book discusses the formulation technologies of phenolipids (lipo-phenolics and pheno-phospholipids). Besides, this book reports on the synthesis of phenolipids, their physicochemical and biological activities, and their potential uses.

Phenolipids as Functional Bioactive Compounds

Phenolics have a low solubility in fats and oils, and improved hydrophobicity by lipophilization was extensively used to render these functionalized compounds. One of the novel techniques used to enhance the antioxidative potential of phenolic compounds is to incorporate lipophilic group(s) to formualte lipophilized phenolics (phenolipids). A promising technique to overcome the low solubility of phenolics is to increase their hydrophobicity by grafting lipophilic moiety to formualte lipid-enriched phenolics (lipo-phenolics). Another method to enhance the amphiphilic traits of phenolics by lipophilization with phospholipids in a suitable solvent to form phenolics-enriched phospholipids (pheno-phospholipids). Both types of lipophilized phenolics (lipo-phenolics and pheno-phospholipids) could be considered phenolipids. However, each phenolipids has particular functional properties and techno-application (Kahveci, Laguerre, & Villeneuve, 2015; Ramadan, 2008, 2012). Phenolipids exhibit higher miscibility in lipo-carriers and lipid phases, offering an advantage for their application in food, drug delivery systems, nutraceuticals, pharmaceuticals, and cosmetics (Kahveci et al., 2015).

An essential factor regarding the antioxidants impact is their localization in the emulsion; however, other important factors such as the interactions with other components could have an effect (Sørensen, Villeneuve, & Jacobsen, 2017). Lipophilization of phenolics is a required field of research to develop antioxidants with enhanced potential. The lipophilization reaction involving the bounding of a lipophilic moieties on a functional groups of a hydrophilic molecules is a novel strategy to control and change the polarity of lipophilized phenolics (Durand et al., 2015). Using lipophilized antioxidants in dispersed lipid models, it has been noted that the antioxidant effect increases with the increase in the chain length up to a critical point, beyond which the effect of lipophilized phenolic decreases. Antioxidant drug designers must seek the significant chain length to formulate the optimal drug (Laguerre et al., 2010; Laguerre, Bayrasy, Panya et al., 2015; Laguerre, Bayrasy, Lecomte et al., 2013). Phenolipids having medium or long chain alkyl tail is novel to expand their applications in lipid-rich products (Aladedunye et al., 2015; Chebil et al., 2007; Hernandez, Chen, Chang, & Huang, 2009; Lorentz et al., 2010; Lue, Guo, Glasius, & Xu, 2010; Ma et al., 2012; Stevenson, Wibisono, Jensen, Stanley, & Cooney, 2006; Zhong & Shahidi, 2011). Gallic acid esters showed different activities with enhanced lipophilicity in emulsions containing phospholipids (Schwarz, 2015). Lipophilization of epigallocatechin gallate (EGCG) with long chain fatty acids resulted in a higher biological potential (Aladedunye et al., 2015; Decker, Warner, Richards, & Shahidi, 2005; Zhong & Shahidi, 2011).

Fig. 1.1 Strategy for phenolipid formation. (Adapted from Figueroa-Espinoza, Laguerre, Villeneuve, & Lecomte, 2013)

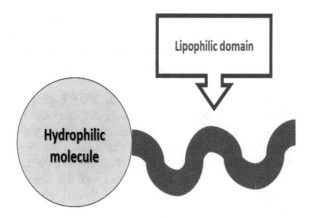

In the biological systems, lipophilization might contribute to easier penetration of antioxidant compounds through the lipid bilayer of cell membranes, that could increase their bioavailability and efficiency (Durand et al., Durand et al., 2017, Durand, Lecomte, & Villeneuve, 2017). The efficient antimicrobial agents interfere with the membrane of the targeted microorganism, wherein antioxidants in cells should have functional cell membranes crossing ability. Phenolics have a limited efficiency because their polarity reduces their ability to penetrate cell membranes. Therefore, new methods that could allow an efficient carrying of active hydrophilic molecules to the specific sites are needed. Phenolipids correspond to a technique where the hydrophilic/lipophilic balance of active hydrophilic molecules is adjusted by covalent grafting of a lipophilic moieties (Fig. 1.1). According to the structure and the reactivity of starting molecules, grafting of the lipid moieties could be carried out using different reactions (esterification, etherification, and transesterification) and catalytic ways (chemical *vs.* enzymatic, and homogeneous *vs.* heterogeneous). The grafted lipophilic moieties could be seen as a key unlocking the lipid barrier encountered by the active compounds. Some phenolipids have exhibited unusual biological activities, wherein some examples were reviewed (Durand, Jacob, et al., 2017; Durand, Lecomte, & Villeneuve, 2017; Figueroa-Espinoza et al., 2013).

Improving the Functional Properties of Phenolics by Lipophilization

It is commonly considered that besides the structural characteristics that affect the antioxidant potential (i.e., position, number, unpaired electron delocalization area, and nature of hydrogen-donating groups), the antioxidant efficiency is governed by its physicochemical traits (polarizability, hydrophobicity, steric hindrance, and diffusivity). The same molecule might be efficient in one system (i.e. oil) but not in another system (e.g., emulsion). Hydrophobicity of antioxidants is of main importance. In the oil-in-water (o/w) emulsions and membranes, apolar antioxidants might locate more at the o/w interface (where the oxidation happens) than their polar analogues, resulting in a higher potential (Frankel & Meyer, 2000). It was anticipated that, in those systems, the more apolar the antioxidant, the higher it's potential.

On the other hand, Laguerre et al. (2009) highlighted the "cut-off" theory, where the relationship between antioxidant activities and hydrophobicity follows a non-linear trend. The antioxidant activity increased with hydrophobicity (concerning alkyl chain length) till it reached a maximum (concerning the chain length), beyond that the lengthening of the alkyl chain led to a collapse in antioxidant impact. The "cut-off" phenomenon was reported with the same series of phenolipids in more complicated systems such as living cells and liposomes where C8-C12 esters were about fivefold more potent than pure phenolic acids. Studies revealed that in hydrophobicity-activity relationships, the "cut-off" effect is considered as the rule rather than the exception. For the same phenolipids series, it was noted that the critical chain length may be different within the tested systems.

Phenolipids are functional compounds displaying several physicochemical and biological traits. This functionality is essential for formulators, as the use of a single bioactive compound with multi-functions reduces the incompatibility between different ingredients. Codex Alimentarius authorized and allowed gallates and parabens as food additives. Propyl, octyl, and dodecyl gallates were considered as antioxidants, and they showed antibacterial and antifungal traits, mostly on Gram-positive bacteria. On the other side, alkyl *p*-hydroxybenzoate and sodium salts are used as antimicrobial agents in food and non-food applications. Some phenolic esters are commercially available and used in cosmetics. Some ferulates are utilized as a skin conditioner, UV filter, or antioxidant. Ethyl caffeate was used as a skin conditioner, while vanillates and salicylates were applied in the flavoring sectors and perfumery (Figueroa-Espinoza et al., 2013; Laguerre et al., 2011; Laguerre, Bayrasy, Lecomte, et al., 2013; Laguerre, Bayrasy, Panya, et al., 2015).

Antioxidants in the International Literature

Thousands of studies have been performed on natural and synthetic antioxidants. A search with the keyword "antioxidant" in Scopus (www.scopus.com) database (May 2020) showed that about 390,000 articles and reviews had been published. Apart from the total published scholarly outputs, about 322,000 were research articles, about 38,000 reviews, and about 7000 book chapters. Figure 1.2 shows the scholarly output on antioxidants since 2000. It is clear that the scholarly outputs published annually on antioxidants are markedly increased from 16,000 contributions in 2010 to 32,000 article in 2019. These measurable indicators reflect the importance of antioxidants as a topic in the international scientific community.

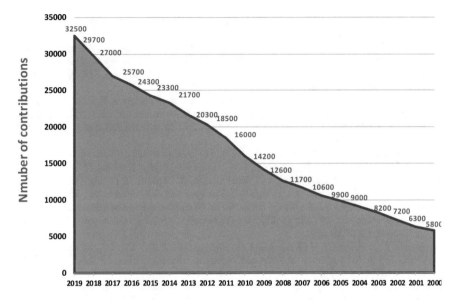

Fig. 1.2 Scholarly output on antioxidants since 2000 (www.scopus.com)

Conclusion and Perspectives

Lipophilization of phenolics is a novel method to synthesize phenolipids that could be utilized in oil-based products, in emulsified, micellar, as well as the liposomal systems. Phenolipids have found industrial applications due to the costs, availability of starting materials, and safety. Recent advances in lipophilization allow accessing molecules with high functionality and targeted action covering a broad spectrum of bioactivities. Further investigations are required to understand the effects of lipophilization on these novel multi-functional compounds, in simple (suspensions, and emulsions) and complecated (cultured cells, and living organisms) systems. Due to their availability and cost, phenolipids should find industrial applications in niche sectors such as pharmaceutics, cosmetics and the novel food. In addition to the functional properties, safety should be an essential factor of success to both consumers and industry.

References

Aladedunye, F., Niehaus, K., Bednarz, H., Thiyam-Hollander, U., Fehling, E., & Matthäus, B. (2015). Enzymatic Lipophilization of phenolic extract from rowanberry (*Sorbus aucuparia*) and evaluation of antioxidative activity in edible oil. *LWT – Food Science and Technology, 60*, 56–62.

Chebil, L., Anthoni, J., Humeau, C., Gerardin, C., Engasser, J., & Ghoul, M. (2007). Enzymatic acylation of flavonoids: Effect of the nature of the substrate, origin of lipase, and operating conditions on conversion yield and regioselectivity. *Journal of Agricultural and Food Chemistry, 55*, 9496–9502.

Decker, E. A., Warner, K., Richards, M. P., & Shahidi, F. (2005). Measuring antioxidant effectiveness in food. *Journal of Agricultural and Food Chemistry, 53,* 4303–4310.

Durand, E., Bayrasy, C., Laguerre, M., Barouh, N., Lecomte, J., Durand, T., et al. (2015). Regioselective synthesis of diacylglycerol rosmarinates and evaluation of their antioxidant activity in fibroblasts. *European Journal of Lipid Science and Technology, 117*(8), 1159–1170. https://doi.org/10.1002/ejlt.201400607

Durand, E., Jacob, R. F., Sherratt, S., Lecomte, J., Baréa, B., Villeneuve, P., et al. (2017). The nonlinear effect of alkyl chain length in the membrane interactions of phenolipids: Evidence by X-ray diffraction analysis. *European Journal of Lipid Science and Technology, 119*(8), 1–7. https://doi.org/10.1002/ejlt.201600397

Durand, E., Lecomte, J., & Villeneuve, P. (2017). The biological and antimicrobial activities of Phenolipids. *Lipid Technology, 29*(7–8), 67–70. https://doi.org/10.1002/lite.201700019

Figueroa-Espinoza, M. C., Laguerre, M., Villeneuve, P., & Lecomte, J. (2013). From phenolics to phenolipids: Optimizing antioxidants in lipid dispersions. *Lipid Technology, 25,* 131–134.

Frankel, E. N., & Meyer, A. S. (2000). The problems of using onedimensional methods to evaluate multifunctional food and biological antioxidants. *Journal of the Science of Food and Agriculture, 80,* 1925–1941.

Hernandez, C. E., Chen, H. H., Chang, C. I., & Huang, T. C. (2009). Direct lipasecatalyzed lipophilization of chlorogenic acid from coffee pulp in supercritical carbon dioxide. *Industrial Crops and Products, 30,* 359–365.

Kahveci, D., Laguerre, M., & Villeneuve, P. (2015). Phenolipids as new antioxidants: Production, activity, and potential applications. *Polar Lipids: Biology, Chemistry and Technology, 5,* 185–214. https://doi.org/10.1016/B978-1-63067-044-3.50011-X

Laguerre, M., Bayrasy, C., Lecomte, J., Chabi, B., Decker, E. A., Wrutniak-Cabello, C., et al. (2013). How to boost antioxidants by lipophilization? *Biochimie, 95*(1), 20–26.

Laguerre, M., Bayrasy, C., Panya, A., Weiss, J., McClements, D. J., Lecomte, J., et al. (2015). What makes good antioxidants in lipid-based systems? The next theories beyond the polar paradox. *Critical Reviews in Food Science and Nutrition.* 55:2, 183-201, https://doi.org/10.1080/10408398.2011.650335

Laguerre, M., Lopez Giraldo, L. J., Lecomte, J., Figueroa-Espinoza, M.-C., Bare'a, B., Weiss, J., et al. (2010). Relationship between hydrophobicity and antioxidant ability of "phenolipids" in emulsion: A parabolic effect of the chain length of rosmarinate esters. *Journal of Agricultural and Food Chemistry, 58*(5), 2869–2876.

Laguerre, M., López Giraldo, L. J., Lecomte, J., Figueroa-Espinoza, M.-C., Baréa, B., et al. (2009). Chain length affects antioxidant properties of chlorogenate esters in emulsion: The cut off theory behind the polar paradox. *Journal of Agricultural and Food Chemistry, 57*(23), 11335–11342.

Laguerre, M., Wrutniak-Cabello, C., Chabi, B., López Giraldo, L. J., Lecomte, J., Villeneuve, P., et al. (2011). Does hydrophobicity always enhance antioxidant drugs? A cut-off effect of the chain length of functionalized chlorogenate esters on ROS-overexpressing fibroblasts. *Journal of Pharmacy and Pharmacology, 63*(4), 531–540. https://doi.org/10.1111/j.2042-7158.2010.01216.x

Lorentz, C., Dulac, A., Pencreac'h, G., Ergan, F., Richomme, P., & Soultani-Vigneron, S. (2010). Lipase-catalyzed synthesis of two new antioxidants: 4-O- and 3-Opalmitoyl chlorogenic acids. *Biotechnology Letters, 32,* 1955–1960.

Lue, B.-M., Guo, Z., Glasius, M., & Xu, X. (2010). Scalable preparation of high purity rutin fatty acid esters. *Journal of the American Oil Chemists' Society, 87,* 55–61.

Ma, X., Yan, R., Yu, S., Lu, Y., Li, Z., & Lu, H. (2012). Enzymatic acylation of isoorientin and isovitexin from bamboo-leaf extracts with fatty acids and antiradical activity of the acylated derivatives. *Journal of Agricultural and Food Chemistry, 60,* 10844–10849.

Ramadan, M. F. (2008). Quercetin increases antioxidant activity of soy lecithin in a triolein model system. *LWT-Food Science and Technology, 41*(4), 581–587. https://doi.org/10.1016/j.lwt.2007.05.008

Ramadan, M. F. (2012). Antioxidant characteristics of phenolipids (quercetin-enriched lecithin) in lipid matrices. *Industrial Crops and Products, 36*(1), 363–369. https://doi.org/10.1016/j.indcrop.2011.10.008

Schwarz, K. (2015). Food antioxidant conjugates and lipophilized derivatives. In *Handbook of Antioxidants for Food Preservation* (pp. 161–176). https://doi.org/10.1016/B978-1-78242-089-7.00007-5

Sørensen, A. D. M., Villeneuve, P., & Jacobsen, C. (2017). Alkyl caffeates as antioxidants in O/W emulsions: Impact of emulsifier type and endogenous tocopherols. *European Journal of Lipid Science and Technology, 119*(6), 1–14. https://doi.org/10.1002/ejlt.201600276

Stevenson, D. E., Wibisono, R., Jensen, D. E., Stanley, R. A., & Cooney, J. M. (2006). Direct acylation of flavonoid glycosides with phenolic acids catalyzed by Candida antarctica lipase B (Novozym 435®). *Enzyme and Microbial Technology, 39*, 1236–1241.

Szydłowska-Czerniak, A., Rabiej, D., & Krzemiński, M. (2018). Synthesis of novel octyl sinapate to enhance antioxidant capacity of rapeseed-linseed oil mixture. *Journal of the Science of Food and Agriculture, 98*(4), 1625–1631. https://doi.org/10.1002/jsfa.8637

Szydłowska-Czerniak, A., Rabiej, D., Kyselka, J., Dragoun, M., & Filip, V. (2018). Antioxidative effect of phenolic acids octyl esters on rapeseed oil stability. *LWT-Food Science and Technology, 96*, 193–198. https://doi.org/10.1016/j.lwt.2018.05.033

Zhong, Y., & Shahidi, F. (2011). Lipophilized epigallocatechin gallate (EGCG) derivatives as novel antioxidants. *Journal of Agricultural and Food Chemistry, 59*, 6526–6533.

Chapter 2
Chemistry, Functionality, and Techno-Applications of Pheno-phospholipids

Abstract Functional phenolics-enriched phospholipids (pheno-phospholipids) play an essential role in enhancing the functional traits of phenolic compounds and phospholipids (PL) in food, nutrition, and health. Phospholipids (PL) are commonly used as emulsifiers and antioxidants. Flavonoids (i.e., quercetin, rutin) exhibit comprehensive biological characteristics, attributable to their antiradical potential. However, the bioavailability of flavonoids, its glycosides, and herbal extracts is essential for its antioxidant potential *in vivo*. The limitation exists and is imputable to poor absorption of flavonoids when orally administered. To overcome this drawback, the development of flavonoids and herbal formulations combined with PL has been produced, which has better absorption and functional traits. Phenolics-enriched lecithin (pheno-phospholipids or phenolipids) and PL-based drug delivery systems (phytosome) are promising as effective herbal drug delivery. Phytosomes showed better therapeutic and pharmacokinetic profiles with novel hepatoprotective and antihepatotoxic properties. Preparation, physicochemical characteristics, and functional properties of pheno-phospholipids and phytosomes are reviewed and discussed in this chapter. The latest developments and the industrial applications of pheno-phospholipids and phytosomes are also described.

Keywords Phospholipids · Phenolipids · Phytosome · Quercetin · Antioxidants · Lecithin · Lipophilic antioxidants

Abbreviations

BHT	Butylated hydroxytoluene
DOPC	Dioleoylphosphatidylcholine
DOPE	Dioleoylphosphatidylethanolamine
DSC	Differential scanning calorimetry
FTIR	Fourier transform infrared
PC	Phosphatidylcholine
PE	Phosphatidylethanolamine

© The Author(s), under exclusive license to Springer Nature Switzerland AG 2021
M. F. Ramadan, *Pheno-phospholipids and Lipo-phenolics*,
https://doi.org/10.1007/978-3-030-67399-4_2

PL   Phospholipids
PUFA   Polyunsaturated fatty acids

Introduction

Lipids are essential food ingredients and one of the main macronutrients required for human nutrition. However, fats and oils are prone to oxidation, which negatively influences food quality, nutritive value, and consumer health. Thus, oxidation is a significant concern to pharma, cosmetics, food, and general public health. This is especially relevant when the lipids contain polyunsaturated fatty acids (PUFA) which are sensitive to oxidation (Cui & Decker, 2016; Damodaran & Parkin, 2008; Ramadan & Asker, 2009). In general, antioxidant efficacy depends on its chemical reactivity (i.e., metal chelator or radical scavenger), interaction with other ingredients, environmental factors (i.e., pH), and the location of antioxidants in the systems (Lucas et al., 2010). For example, the environment polarity affects the antioxidant traits of phenolic compounds (Reddy, Shanker, Ravinder, Prasad, & Kanjilal, 2010). The polar paradox theory (Lucas et al., 2010; Shahidi & Zhong, 2011) exhibits the paradoxical behavior of antioxidants in different media and rationalizes that non-polar antioxidants are more effective in more polar media (i.e., liposomes and oil-in-water emulsions), while polar antioxidants are more effective in less polar media (i.e., oils and fats).

Phospholipid Characteristics and Functional Properties

Phospholipids (PL) are constituents of cell membranes and found in many food-stuffs from both plant and animal sources (Cui & Decker, 2016). Phospholipids possessing functional traits such as emulsification, wetting agent, anti-spattering, crystallization inhibition, and non-stick release are commonly added to foodstuffs (Cui & Decker, 2016; Van Nieuwenhuyzen & Tomás, 2008). Lecithin (crude PL) is a mixture of PL including phosphatidylcholine (PC), phosphatidylethanolamine (PE) and phosphatidylinositol, and considered as an essential source of choline for nutritional formulations. Lecithin recovered from vegetable oil refining contains about 40% neutral lipids (triacylglycerols), and the remainder part contains polar lipids (glycolipids and PL). Lecithin serves as an emulsifying agent in caramel, chocolate, margarine, and chewing gum. Some infant formulas, baked goods, and dairy products contain lecithin as a wetting agent. The levels and composition of PL vary in foodstuffs that depend on the food origin and processing technique (Cui & Decker, 2016; Rombaut & Dewettinck, 2006). Non-food uses of lecithin include paper, polishes, plastic, waxes, paints, wood coating, magnetic-type media, printing, cosmetics, and pharmaceuticals (Joshi, Paratkar, & Thorat, 2006; Ramadan, 2008; Ramadan & Asker, 2009).

The molecular structure of PL could be changed enzymatically or chemically (Vikbjerg, Rusig, Jonsson, Mu, & Xu, 2006). Antioxidant and pro-oxidant mechanisms of PL were reviewed for food applications (Cui & Decker, 2016). Antioxidant potentials of PL were studied through their addition to edible fats and vegetable oils (King, Boyd, & Sheldon, 1992; Ramadan, 2012; Ramadan & Asker, 2009). Antioxidative characteristics of PL have been claimed due to (a) synergism between natural antioxidants, tocopherols and PL (Judde, Villeneuve, Rossignol-Castera, & le Guillou, 2003); (b) induction of Maillard-reaction products between amino PL and oxides (Husain, Terao, & Matsushita, 1984); (c) chelating pro-oxidant metals by phosphate group (Jewell & Nawar, 1980); and (d) effect as an oxygen barrier between air and oil interfaces (Porter, 1980).

PUFA in PL are the main sensitive sites for oxidation. Therefore, PL are easily oxidized (Passi et al., 2004), wherein the oxidative process modifies PL characteristics responsible for their antioxidant potential (Ramadan, 2008, 2012). The cability of PL to interact with oil and water makes it a novel emulsifiers (Vikbjerg et al., 2006). Phospholipids have been used in oil-in-water (o/w) and water-in-oil (w/o) emulsions, in the processing of food, and pharmaceuticals. Phospholipids are amphiphilic compounds contain hydrophilic heads and hydrophobic tails, being able to rearrange themselves as a liposome pherical and closed structure containing lipid bilayer (Bouarab et al., 2014; Gonçalves et al., 2015; Liu, Du, Zeng, Chen, & Niu, 2009; Ramadan, 2008). Liposomes and phytosomes are novel carriers for the delivery of hydrophobic/hydrophilic molecules and have an affinity to cell membranes that make it able of increasing drug absorption (Gonçalves et al., 2015; Hoeller, Sperger, & Valenta, 2009).

Phenolics Bioavailability, Characteristics, and Functional Properties

Phenolics have gained interest due to their health-promoting and biological traits (Kroll, Rawel, & Rohn, 2003). Flavonoids exhibit antioxidant potential in biosystems, but their functional properties depends on their chemical characteristics (Rice-Evans & Miller, 1996; van Acker et al., 1996). Quercetin (Fig. 2.1) is a

Fig. 2.1 Quercetin (2-(3,4-dihydroxyphenyl)-3,5,7-trihydroxy-4H-chromen-4-one)

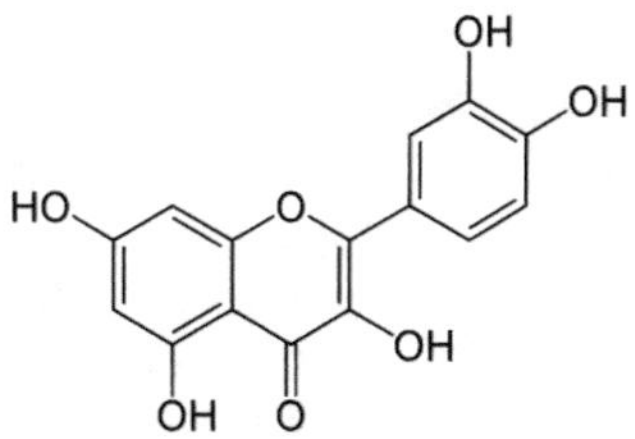

flavonoid found in onion, tea, apple, and berries as aglycone. Quercetin exhibited numerous health-promoting, antiinflammatory, anticancer, antiallergic impacts and enhanced cardiovascular health (Maiti et al., 2005; Makris & Rossiter, 2001; Terao & Piskula, 1999). Besides, quercetin could quench reactive oxygen species (ROS), and down-regulate lipid oxidation (Das et al., 2008; Lévai, Martín, De Paz, Rodríguez-Rojo, & Cocero, 2015). The catechol group in quercetin contributes directly to the chelating effect (Brown, Khodr, Hider, & Rice-Evans, 1998; Lévai et al., 2015; Maiti et al., 2005; Parmara, Singhb, Bahadurc, Marangonib, & Bahadura, 2011; Ramadan, 2012).

Several factors could affect the absorption of flavonoids, including glycosylation on hydroxyl group, the position of glycosylation, attached sugar moiety, food/plant medium, and interaction with proteins, micelles and emulsifiers (Nemeth & Piskula, 2007). Flavonoid glycosides are hardly absorbed in the small intestine due to the sugar moiety that elevates their hydrophilicity. The other factors for poor absorption are the bacterial degradation of the phenol moiety and a complex production with other substances in the gastrointestinal tract (Maiti et al., 2005).

There is limited data about *in vivo* bioavailability of flavonoid glycosides and quercetin. The compound's solubility influences the efficiency of quercetin intestinal absorption in the vehicles. In the grape juice, the absorption of quercetin glycosides was less than from the aglycons (Meng, Maliakal, Hong, Lee, & Yang, 2004). Traces of quercetin was detected in the plasma after the ingestion of grape juice or pure aglycon (Vitaglione, Morisco, Caporaso, & Fogliano, 2005). Low bioavailability is the main limitation of quercetin applications, making it essential to be administered in high amounts (Somsuta et al., 2012). Because of the poor solubility in the gastrointestinal tract, quercetin has minimal absorption, and its bioavailability was very low in the experimental animals (Khaled, El-Sayed, & Al-Hadiya, 2003; Maiti, Mukherjee, Gantait, Saha, & Mukherjee, 2006; Murray, Booth, Deeds, & Jones, 1954) and humans (Gugler, Leschik, & Dengler, 1975; Russo, Spagnuolo, Tedesco, Bilotto, & Russo, 2012). The low solubility limits quercetin's bioactivities *in vivo* (Gonçalves et al., 2015; Souza et al., 2013; Srinivas, King, Howard, & Monrad, 2010). On the other hand, the effectiveness of herbal drugs depends on delivering a sufficient level of bioactive compounds (Maiti et al., 2005; Mukherjee, 2001, 2002). Therefore, it was essential to develop flavonoids' preparations capable to enhance its solubility which resulting in an increase *in vivo* bioavailability (Gonçalves et al., 2015).

Different approaches have been suggested and discussed to increase quercetin bioavailability. Azuma, Ippoushi, Ito, Higashio, and Terao (2002) enhanced quercetin absorption by combing lipids and emulsifiers. Thus, quercetin dispersion in lipomicelles might be an excellent factor for its better absorption in the tract (Azuma et al., 2002; Nemeth & Piskula, 2007). Besides, catechins from green tea were better absorbed when supplemented as a PL-complex. Some companies developed and marketed complexes of PL with herbal extracts (Phytosome) to improve bioavailability. Mulholland et al. (2001) prepared water-soluble quercetin, and its availability was 20%. To increase the availability of quercetin, drug-loaded lipid nanoparticles might be an alternative, wherein the complexation of quercetin with cyclodextrin and lecithin in aqueous solution was studied (Pralhad & Rajendrakumar, 2004; Yuan et al., 2006). Li et al. (2009) prepared by emulsification lecithin-encapsulated quercetin. The absorption levels of quercetin-loaded lipid nanoparticles was tested, obtaining a six-fold increase in the bioavailability, compared to pure quercetin (Lévai et al., 2015).

Phenolics-Enriched Phospholipids (Pheno-phospholipids)

Most of the flavonoids and active constituents from herbal drugs are poorly absorbed due to their multiple-ring large size and to their low miscibility with lipid, severely limiting their capability to pass through the lipid-rich membranes of the enterocytes of the small intestine (Manach, Scalbert, & Morand, 2004; Semalty, Semalty, & Rawat, 2007). Water-soluble phenolics could be modified to lipid-compatible complexes, which called "Phytosomes". The term "Phyto" means plant, and "some" means cell-like (Citernesi & Sciacchitano, 1995; Semalty et al., 2007). The phytosome process was applied to phyto-extracts, including grape seed, hawthorn, milk thistle (*Silybum marianum*), *Ginkgo biloba*, ginseng (*Panax ginseng*), and green tea (*Thea sinensis*). The flavonoids and terpenoids of phyto-extracts lend themselves well for the binding to PC. Phytosomes are more bioavailable compared to phyto-extracts because of their enhanced capacity to cross the lipid-rich membranes and reach the blood (Bombardelli et al., 1989; Semalty et al., 2007). The lipid-phase constituents employed to make phytochemicals lipid-compatible, are PL. PC, the main compound building block of cell membranes, is miscible in oil/lipid and in water environments, and orally well absorbed.

There are some approaches to increase the availability of quercetin, whether by the development of colloidal quercetin delivery systems or by chemical modifications (Barras et al., 2009). A controlled release of quercetin by its encapsulation into L-lactide (PLA) and poly-D nanoparticles *via* the solvent evaporation method was achieved (Gonçalves et al., 2015; Kumari, Yadav, Pakade, Singh, & Yadav, 2010). Wu et al. (2008) prepared quercetin-loaded nanoparticles using a nano-precipitation method using polyvinyl alcohol and Eudragit® E as carriers, thus obtaining a quercetin's release 74-fold more in comparison with the quercetin. In addition, quercetin was encapsulated in Pluronic F127 through the supercritical anti-solvent technique, enabling better dissolution behavior quercetin in simulated physiological fluids (Fraile, Buratto, Gomez, Martín, & Cocero, 2014). Althans, Schrader, and Enders (2014) used hydrogels and hyperbranched polymers as quercetin delivery systems to increase its aqueous solubility and to improve the stabilization of flavonoid. Cyclodextrins have also been used to encapsulate quercetin using the co-evaporation or freeze-drying techniques (Calabro et al., 2004; Gonçalves et al., 2015; Pralhad & Rajendrakumar, 2004).

Barras et al. (2009) applied lipids to encapsulate quercetin and to increase its apparent water solubility. In addition, nanostructured lipid carriers, solid-lipid nanoparticles, and lipid nanoemulsions were prepared, achieving about 90%

encapsulation efficiencies, using a high-pressure homogenizer for quercetin encapsulation (Aditya et al., 2014). Quercetin-loaded liposomes were formulated (Gonçalves et al., 2015; Mignet, Seguin, & Chabot, 2013; Priprem, Watanatorn, Sutthiparinyanont, Phachonpai, & Muchimapura, 2008). Inorganic materials could be used as a carrier for the encapsulation of quercetin, wherein quercetin-loaded silica microspheres were formulated using polyol-in-oil-in-water emulsion and solegel techniques to enhance the stability of flavonoid (Gonçalves et al., 2015; Kim et al., 2015). De Paz et al. (2012; De Paz, Martín, & Cocero, 2012) developed a technique for the encapsulation of hydrophobic compounds, based on ethyl acetate-water emulsions' formualtion at high temperature and pressure. A stable water suspension of β-carotene with micellar particle sizes (*ca.* 400 nm) and encapsulation efficiencies (*ca.* 80%) was achieved.

Bandarra, Campos, Batista, Nunes, and Empis (1999) mentioned the ability of lecithin-containing high levels of PE, PC, and cardiolipin combinations to increase tocopherol activity. In general, PC has not been reported to enhance the activity of tocopherols. One suggested mechanism depend on the cooperative action of PL and tocopherols might be due to the head group of PE and PC regenerating tocopheroxyl radical to tocopherol (Cui, Mcclements, & Decker, 2015). Doert, Jaworska, Moersel, and Kroh (2012) revealed that primary amines could restore oxidized tocopherols *via* an ionic mechanism rather than a radical one. A second suggested mechanism for PL improving the activity of tocopherols is the non-enzymatic browning reaction products from amine-containing PL, as well as the oxidation products such as aldehydes are antioxidants and therefore provide other sources of antioxidants (Chen, Han, Laguerre, McClements, & Decker, 2011; Cui et al., 2015; Laguerre, Lecomte, & Villeneuve, 2015). Doert, Krüger, Morlock, and Kroh (2017) also reported the synergistic impact of lecithin for tocols and the antioxidant potential of the PE-ascorbic acid condensate.

Phenolics conjugated with PL were investigated, but few studies were published on this topic. Maiti et al. (2005) formulated a quercetin-PL complex with better absorption and functional properties. The antiradical activity of quercetin-PL complex and pure quercetin was evaluated in oxidative stress conditions in rats induced by CCl4 intoxication. Preparation of PL and phenolics containing quercetin-lecithin as a non-covalent mixture showed an increase in the antioxidant properties with increasing the concentration of quercetin-lecithin complex (Balakrishna et al., 2017; Ramadan, 2008; Ramadan, 2012). Quercetin-enriched lecithin mixture showed higher antimicrobial traits compared to pure lecithin and quercetin, which reveals that the lipidic part synergistically enhanced the bioactivity (Ramadan, 2008; Ramadan & Asker, 2009).

Liposomes are prepared by mixing PL with water-soluble substances without chemical bonds (Fig. 2.2). Pheno-phospholipids unlike liposomes, induced from the reaction of PL with phenolics (Ramadan, 2013, 2014). It was observed that the formulation did not contain covalent lipo-conjugates of phenolics. Therefore, pheno-lipid conjugates based on PL could have novel uses and applications compared with non-covalent complex. In this regard, the synthesis of functionalized PL with

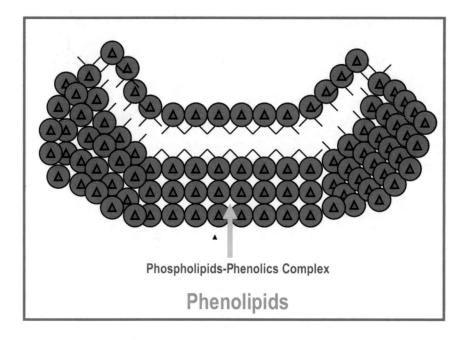

Phospholipids-Phenolics Complex

Phenolipids

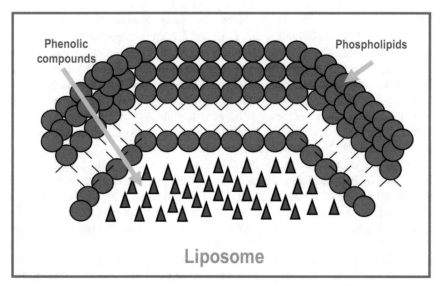

Phenolic compounds

Phospholipids

Liposome

Fig. 2.2 Structural differences between pheno-phospholipids and liposomes

phenolics would be interesting because PL are amphiphilic constituents that can expand the use of phenolics in both pharmaceutical and food industry.

PC have several applications in the food production due to its emulsifying properties as an anti-staling agent, an anti-spattering agent, and due to its pharmaceutical

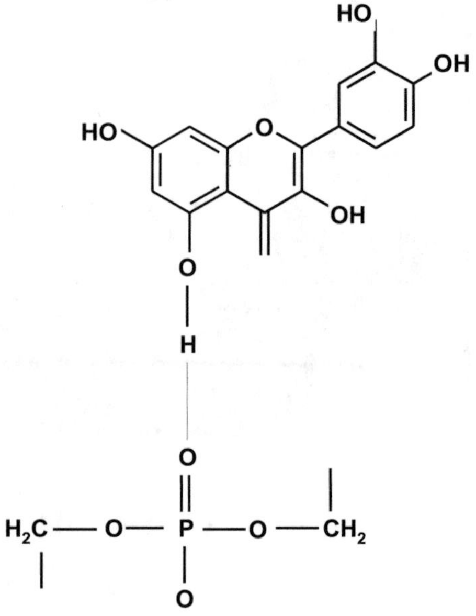

Fig. 2.3 Quercetin-PC
complex

features (Balakrishna et al., 2017; Kiełbowicz, Gładkowski, Chojnacka, & Wawrzenczyk, 2012; Kim et al., 2009; Selmair & Koehle, 2009). Nasibullin, Nikitina, Afanas'eva, Nasibullin, and Spirikhin (2002) investigated the interaction of PC with quercetin by quantum chemistry and NMR. They reported the capability of PC and quercetin to form chain structure linked by a hydrogen bond (Fig. 2.3). Ramadan (2008) and Ramadan and Asker (2009) studied the changes produced antioxidant and antimicrobial activities of PL following complexing with quercetin. The antioxidant characteristics of the pure lecithin and complexes (quercetin-enriched lecithin) when added to triolein models and sunflower oil during accelerated thermal oxidation test, was examined (Ramadan, 2012).

Dihydromyricetin is a natural antioxidant, but its weak lipophilic properties limits its use in lipophilic foodstuffs. The dihydromyricetin-lecithin complex was formulated to increase the hydrophobicity of dihydromyricetin. A curcumin-encapsulated o/w microemulsion system was prepared using lecithin and ethyl oleate as the oil phase and Tween 80 as the surfactants. The curcumin in micro-emulsion was stable for 2 months (Lin, Lin, Chen, Yu, & Lee, 2009). Balakrishna et al. (2017) synthesized structured-PC containing phenolics and examined their antioxidant and antimicrobial traits. The structure of phenoylated-PC was confirmed by FTIR, NMR, and mass spectral analyses. *in vitro* antimicrobial and antioxidant effects of the formulations were evaluated. 1-glutaryl-phloroglucinol-2-docosahexaenoyl PC was constructed and tested for its impact against all-*trans*-retinal toxicity (Balakrishna et al., 2017; Crauste et al., 2014). In another report, PC was enzymatically constructed with ethyl ferulate to obtain feruloylated lyso PC (Yang, Mu,

Chen, Xiu, & Yang, 2013). The resulted products might be non-toxic and envisaged to have novel uses and applications as natural emulsifiers in addition to antioxidant traits. The presence of PC further improved the biological traits of phenolic acids (Balakrishna et al., 2017).

Preparation of Pheno-phospholipids

Preparation of Phytosomes

Jiang, Yu, Yan, and Chen (2001), and Maiti et al. (2006) studied the techniques of phytosome preparation. Complex phenolics formulated phytosomes in 1:2 or 1:1 ratio with PC. As part of the phytosome process, silybin was isolated from the other molecules (silydianin and silychristin) for the most significant effects. The molecular binding to a compound that is readily absorbed is what increases the absorption of silybin in Milk Thistle phytosome up to 10 times more than control milk thistle (Semalty et al., 2007). Yanyu, Yunmei, Zhipeng, and Quineng (2006) constructed, using ethanol as a reaction medium, a silybin-PL complex. Silybin and PL were dissolved into ethanol, and the solvent was then removed under vacuum to obtain a silybin-PL complex. Naik and Panda (2007) prepared *Ginko* phytosomes by reacting to a stoichiometric amount of PL with *Ginkgo biloba* extract.

Gonçalves et al. (2015) developed water-soluble quercetin preparations through pressurized ethyl acetate-water emulsion method. They formulated quercetin using natural surfactants (i.e., lecithin, OSA-starch, and β-glucan) by precipitation from a pressurized ethyl acetate in-water emulsion. The effects of the process parameters were studied, including quercetin and emulsifier's concentrations, suspension of quercetin and dissolution of emulsifier, flows of solvent, and the organic to water ratio. Using lecithin, the production of quercetin with encapsulation efficiency up to 76% and micellar particle size in the range of nanometers were obtained. An enhanced antioxidant potential (three-fold higher) was reported in these preparations, revealing that lecithin is a novel emulsifier for quercetin encapsulation.

Encapsulated quercetin was produced on a sub-micrometric scale to increase its availability. Particles were prepared by extraction of organic solvent from o/w emulsions by supercritical fluid extraction (SFE) of the emulsions. Due to the fast extraction of the solvent, the organic phase becomes supersaturated, resulting in quercetin precipitation in the sub-micrometric scale, encapsulated by the surfactant materials. Two biopolymers (lecithin and Pluronic L64® poloxamers) were utilized as carriers. Quercetin-loaded multi-vesicular liposomes were prepared, with 70% quercetin encapsulation efficiency and 100 nm particle size, without segregated quercetin crystals (Lévai et al., 2015).

Preparation of Dihydromyricetin-Lecithin Complexes

Lecithin and dihydromyricetin were dissolved and stirred in tetrahydrofuran. After removing the solvent, the residue was collected and ground. The resultant yellow power was received as a dihydromyricetin-lecithin complex. A physical mixture of lecithin and dihydromyricetin was also prepared by mixing lecithin and dihydromyric-etin at room temperature, and the mixture was stirred. The recovered product was collected as a dihydromyricetin-lecithin complex physical mixture (Liu et al., 2009).

Preparation of Phenolic Acids-Lecithin Complexes

Balakrishna et al. (2017) synthesized phenoylated-PC from egg PC and phenolic acids (i.e., sinapic, ferulic, vanillic, and syringic). The structures of phenoylated-PC were tested by FTIR, NMR, and mass spectral analyses. 2-palmitoyl lyso PC was prepared from PC *via* regioselective enzymatic hydrolysis and reacted with hydroxyl-protected phenolic acids to form phenoylated-PC. The synthetic mecha-nisms are as shown in Schemes 2.1, 2.2, and 2.3. Hydroxyl groups of phenolic acids were protected with *tert*-butyldiphenylsillylchloride in the presence of imidazole to form protected phenolic acids (Scheme 2.1). The 2-palmitoyl and 2-acyl lyso PC were purified and acylated with protected phenolic acids (i.e., benzoic acid and cinnamic acid derivatives) to produce corresponding phenolic modified PC. The TBDPS group's deprotection was carried out to obtain 1-phenoyl-2-acyl PC (Schemes 2.2 and 2.3).

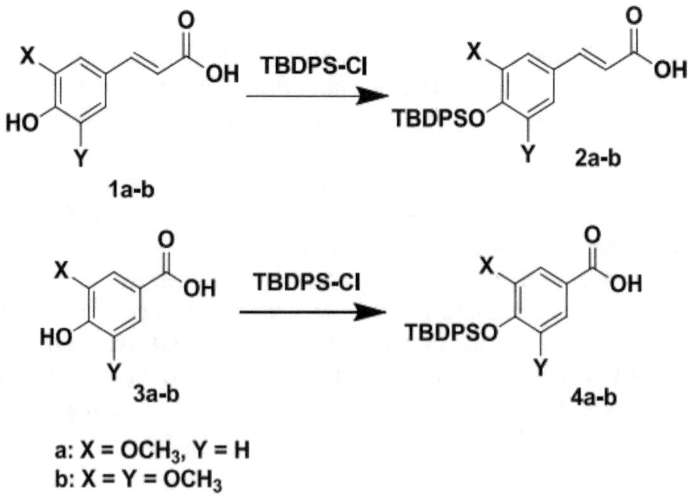

a: X = OCH_3, Y = H
b: X = Y = OCH_3

Scheme 2.1 Synthesis of protected phenolic acids (Balakrishna et al., 2017)

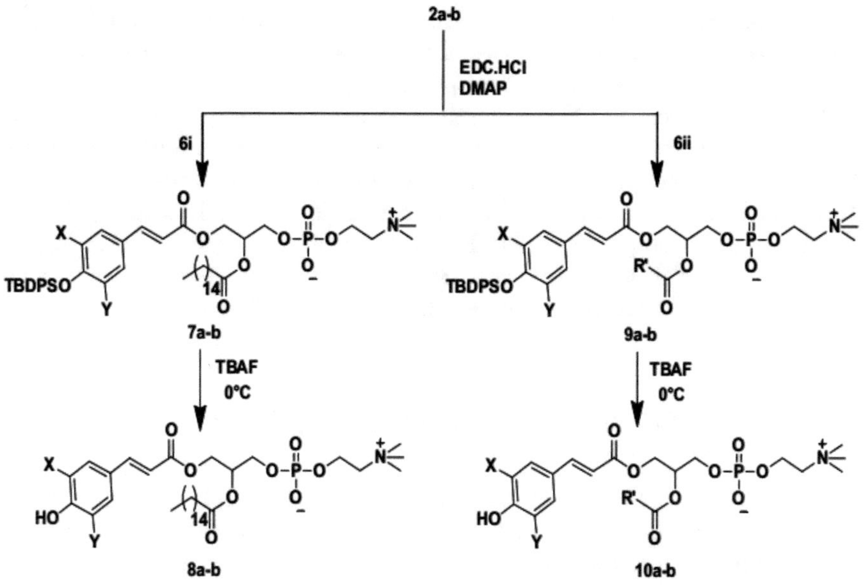

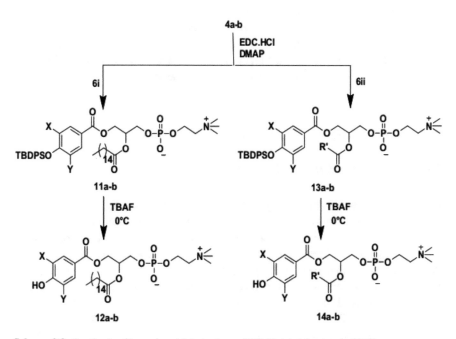

Scheme 2.2 Synthesis of cinnamic acid derivatives of PC (Balakrishna et al., 2017)

Scheme 2.3 Synthesis of benzoic acid derivatives of PC (Balakrishna et al., 2017)

Anankanbil, Pérez, Banerjee, and Guo (2018) prepared sn-1-acyl (C12–C18)-sn-2-caffeoyl, and sn-1-caffeoyl-sn-2-acyl phosphatidylcholines using mild scalable regiospecific pathways. Phenophospholipids showed superior emulsion stability than soybean PC. Phenophospholipids exhibited superior oxidation inhibition, while incorporation of caffeoyl in PC did not affect the antiradical ability of caffeic acid. Recently, Okulus and Gliszczyńska (2020) investigated lipase-catalyzed acidolysis reactions of PC with veratric and anisic acids to prepare O-methylated pheno-phospholipids. For the incorporation of anisic acid in PC, the most active biocatalyst was Novozym 435. In addition, the mixture of toluene: chloroform (9:1, v/v) increased the incorporation of anisic acid into PC.

Preparation of Quercetin-Phospholipid Complexes

Complexes of quercetin with PL were prepared (Bombardelli & Patri, 1991; Maiti et al., 2005). One mole of quercetin was refluxed with 1 mole of PC in dichloromethane till the quercetin dissolved. The resulting solution's volume was reduced, and n-hexane was added to get the complex as a precipitate. The mixture was filtered and dried under vacuum (Maiti et al., 2005). Ramadan (2008, 2012) prepared quercetin-enriched lecithin by mixing quercetin with lecithin (1:1, w/w) then the complex was dissolved at 40 °C in ethyl acetate. With low toxicity, ethyl acetate is a Generally Recognized as Safe (GRAS) solvent, and it might be safely used as a flavoring agent (Riemenschneider, 2000). The mixture was concentrated and freeze-dried to obtain quercetin-enriched lecithin complex. Patent has been published concerning the chemical synthesis of lipo-phenolics (Hassanien, 2016).

Characterization and Functionality of Phenolipids

Functional Properties of Phytosomes

Phytosomes produced from the reaction of PL with phenolics (i.e., flavonoids) in a non-polar solvent (Bombardelli et al., 1989). Phytosomes are lipophilic compounds with a specific melting point, moderately soluble in fats and freely soluble in non-polar solvents. When treated with water, phytosomes assume a micellar shape, producing structures that resemble liposomes. In liposomes, the active principle is dissolved in the layers of the membrane or the medium of the cavity. In contrast, in phytosomes, it is an integral part of the membrane. Molecules are anchored via chemical bonds to PL polar head (Bombardelli, 1991; Bombardelli & Spelta, 1991). A liposome is constructed by mixing lecithin and a water-soluble substance. No chemical bond is formed, wherein the PC molecule surround the water-soluble material. In contrast, in phytosomes, lecithin and phenolics form complexes which are better absorbed than liposomes (Semalty et al., 2007).

Phytosomes have the following merits (Bombardelli, 1994; Bombardelli, Spelta, Loggia Della, Sosa, & Tubaro, 1991; Kidd & Head, 2005; Semalty et al., 2007).

- Phytosomes improve the absorption and bioavailability of phenolics.
- As the absorption and bioavailability of phenolics are improved, its dose requirement could be reduced.
- Lecithin (i.e., PC), used in phytosomes preparation, acts as a hepatoprotective agent.
- Chemical bonds are formed between PC and phenolics; thus, the phytosomes showed improved stability.

Functional Properties of Silymarin-Phytosomes and Silipide

Hepatoprotective potential of silymarin-phytosomes and silipide (a complex of silybin with PC) were reviewed (Semalty et al., 2007). Basaga, Poli, Tekkaya, and Aras (1997) reported the antioxidant properties of silibin-PC complex and its ability to react with the superoxide radical anion, and the hydroxyl radical. Giacomelli et al. (2002) mentioned that silybin-PL complex potentiates *in vitro* and *in vivo* activities of cisplatin. Filburn, Kettenacker, and Griffin (2007) reported the enhanced bioavailability of a silybin-PC complex in dogs. Provinciali et al. (2007) revealed the antitumor impact of the silybin-PC complex on the mammary tumors in HER-2/neu mice. The antitumor impacts were attributed to the down-regulation of HER-2/neu-expression and the formation of senescent-like growth arrest and apoptosis in tumor cells through a p53-mediated pathway. Naik and Panda (2007) revealed the protective effects of *Ginkgo biloba* phytosomes on CCl_4-induced hepatotoxicity in the experimental animals due to its antioxidant and antiradical activity. Falasca et al. (2008) highlighted the hepatoprotective, antifibrotic, and anti-inflammatory effects of silybin-PL and vitamin E complex.

Functional Properties of Dihydromyricetin-Lecithin Complexes

Liu et al. (2009) prepared dihydromyricetin-lecithin complex in tetrahydrofuran and a physical mixture of dihydromyricetin with lecithin. Lecithin and dihydromyricetin in the complex were combined by non-covalent bonds and did not form a new compound. The lipophilic property of dihydromyricetin and the solubility of dihydromyricetin in *n*-octanol were improved. The lecithin-dihydromyricetin complex was an effective quencher of DPPH· free radicals. Using lard oil as a substrate in the Rancimat test, the performance of lecithin-dihydromyricetin complex with a protection factor of 6.67 was superior to that of BHT. The DSC curve of dihydromyricetin exhibited 3 endothermal peaks. The first two peaks might be due to the removal of two crystal water from dihydromyricetin, while the third peak was due to dihydromyricetin melting. DSC curve of the physical mixture exhibited the impact of dihydromyricetin. Still, the DSC curve of the lecithin-dihydromyricetin showed the effect of lecithin, in which the characteristic peaks of dihydromyricetin werevdisappeared. The unique absorption peaks of dihydromyricetin were recorded at 290 nm. The UV spectra showed no differences between the complex and the physical mixture. The IR spectra showed no significant differences between the complex and the physical combination. The spectra of the complex and the physical mixture showed an additive effect of dihydromyricetin and lecithin and. The characteristic absorption peaks of lecithin and dihydromyricetin were recorded at 1742 cm^{-1} and 1641 cm^{-1}, respectively. However, in the spectrum of their complex, some small characteristic absorption peaks of dihydromyricetin between 1000 cm^{-1} and 1500 cm^{-1} were masked by lecithin. In IR analysis, no new peaks were recorded in the complex and mixture. These findings suggested that some weak physical interactions between lecithin and dihydromyricetin induced during the complex formation.

Functional Properties of Quercetin-Phospholipid Complexes

Maiti et al. (2005) formulated a quercetin-PL complex with a higher absorption trait. The antiradical activity of quercetin-PL complex and pure quercetin was tested under oxidative stress conditions in rats induced by CCl$_4$. Quercetin-PL complex restored the liver glutathione system's reduced enzyme levels as well as impaired levels of other enzymes, which were significant concerning the CCl$_4$-treated group. The complex showed better effects than pure quercetin at the same doses. The results ascertain the superiority of quercetin-PL complex over pure quercetin in terms of better antiradical action.

Gonçalves et al. (2015) formulated quercetin using lecithin by precipitation from a pressurized ethyl acetate in-water emulsion. Since the emulsifier has a critical micelle concentration (20 g/L), an equal or higher level of lecithin was utilized to ensure the micelle formation (De Paz, Martín, & Cocero, 2012; Varona, Martín, & Cocero, 2009). The test carried out with organic-water ratio and lecithin's concentration at higher levels led to the production of an unstable emulsions. The emulsions were more stable at lower solvent: water ratio and lower lecithin level. None of the emulsions exhibited the presence of quercetin's crystals, revealing a good quercetin encapsulation. Crystals of quercetin were precipitated during the evaporation of ethyl acetate, revealing that quercetin's concentration of 7.5 and 10 g/L might be high for its complete dissolution in the pressurized hot organic solvent, and thus high for good encapsulation. In addition, the formation of crystals was tested by optical microscopy (Fig. 2.4). Figure 2.5 presents the Cryo-TEM figures of the suspension. By studying the Cryo-TEM figures of lecithin's water suspension (Fig. 2.5a), it could be observed multivesicular and multilamellar vesicular formulations of the emulsifier with 100 nm particle size (Varona et al., 2009). Concerning the quercetin aqueous suspensions, the presence of small vesicles with dark double

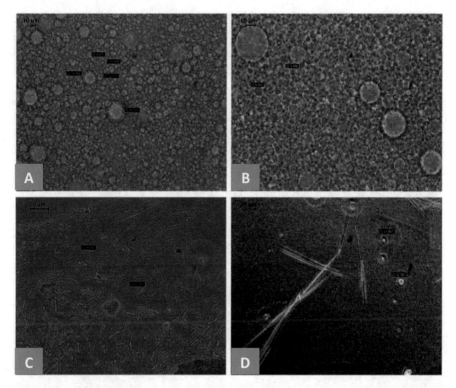

Fig. 2.4 Microscopic images of quercetin emulsions (**a** and **b**) and suspensions (**c** and **d**) stabilized with lecithin (Gonçalves et al., 2015)

layers could be noted (Fig. 2.5b, c) possibly due to the presence of quercetin. In Fig. 2.5d, larg vesicles with the beehive-like structures were noted, possibly due to the formation of quercetin-enriched lecithin structure linked by H bond. Lecithin could be considered as an excellent emulsifier for the encapsulation of quercetin. The capability of these two compounds to construct a chain-like structure linked by H bonds has been confirmed with nuclear magnetic resonance (NMR) spectroscopy (Nasibullin et al., 2002). Furthermore, a synergism between quercetin and lecithin and seems to be found due to the formation of these bonds, resulting in antioxidant activities (Ramadan, 2012).

Fourier transform infrared (FTIR) spectroscopy studied the production of a complex or chemical association between quercetin and lecithin (Fraile et al., 2014). In Fig. 2.6, FTIR spectrum of unprocessed quercetin hydrate shows its unique bands, the broadband at 3500–3000 cm^{-1}, assigned to a free -OH bond vibration, the bands at 1310 cm^{-1} and 1160 cm^{-1}, assigned to the C-O-C vibration, the band at 1515 cm^{-1}, assigned to aromatic groups, bands at 1660 cm^{-1} and 1600 cm^{-1}, assigned to the stretching vibration of the C=O group, and the band at 1010 cm^{-1}, assigned to aromatic C-H groups (Fraile et al., 2014; Yang et al., 2013). Regarding FTIR spectra of the encapsulated quercetin, some complexation between quercetin and lecithin were

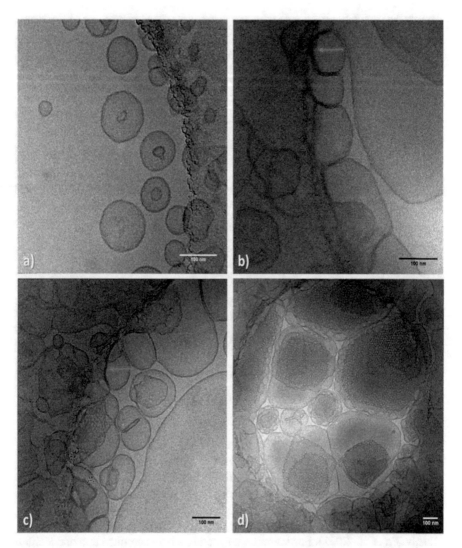

Fig. 2.5 TEM images of (**a**) lecithin water suspensions and (**b–d**) encapsulated quercetin suspensions. Orange arrows show the double dark layers (Gonçalves et al., 2015)

noted, since it presents different band shapes recorded between 1500 cm^{-1} and 1660 cm^{-1} when compared with unprocessed physical mixture and quercetin spectra. Quercetin's crystals showed a unique carbonyl absorption band of around 1600 cm^{-1}, which was assigned to aromatic ketonic carbonyl stretching (Pralhad & Rajendrakumar, 2004; Zheng & Chow, 2009), that is not detected in the spectrum of encapsulated quercetin, revealing a good encapsulation of quercetin.

Ramadan (2008) and Ramadan (2012) studied the antioxidant potential of pure lecithin and mixtures of lecithin and quercetin (1:1, w/w) in the protection of triolein

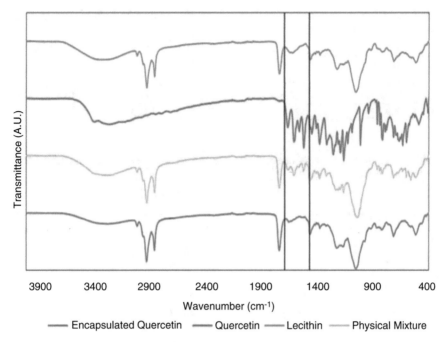

Fig. 2.6 FTIR spectra of encapsulated quercetin, quercetin and lecithin, and a physical mixture of quercetin: lecithin (1:75). Black lines represent the spectrum range characteristic of crystalline quercetin (Gonçalves et al., 2015)

and sunflower oil subjected to heat conditions (60 °C). Absorptivity at 232 nm and 270 nm in samples contain lecithin increased during the test, due to the formation of conjugated dienes (CD) and conjugated trienes (CT). Oxidative stability of quercetin-lecithin enriched triolein samples were higher than those containing pure quercetin or lecithin, possibly due to the synergism between quercetin and lecithin. Furthermore, the increase in the level of the quercetin-lecithin complex increased its antioxidative activity. Ramadan and Asker (2009) studied the antimicrobial and antiviral impacts of pure lecithin, pure quercetin, and quercetin-enriched lecithin. The antimicrobial and antiviral traits of quercetin-enriched lecithin complexes were higher than pure quercetin or lecithin (Tables 2.1 and 2.2). Quercetin-enriched lecithin complexes exhibited intense antimicrobial action in comparison with pure lecithin and pure quercetin. Quercetin-enriched lecithin complex (1:99, w/w) exhibited high antibacterial properties against gram-positive bacteria with a minimum inhibitory concentration (MIC) between 750 μg/mL and 1000 μg/mL. Furthermore, quercetin-enriched lecithin (3:97, w/w) showed a drastic impact on the biosynthesis of protein in the cells of *B. subtilis*. Still, the impact was slightly observed on the biosynthesis of RNA and DNA. Quercetin-enriched lecithin complexes could be applied to enhance functionality and health effects of lecithin and flavonoids in novel food and pharmaceutical products.

Table 2.1 Antibacterial activity (inhibition zone, mm) of quercetin, lecithin, and quercetin-enriched lecithin (Ramadan & Asker, 2009)

Compound/formulation	*P. aeruginosa*	*E. coli*	*St. aureus*	*B. subtilis*
Quercetin (Qu)	13.0	15.6	15.3	19.0
Lecithin (Le)	10.3	9.6	8.33	9.00
Qu-Le (1:99, w/w)	17.6	18.6	17.0	19.6
Qu-Le (3:97, w/w)	24.3	23.3	22.0	27.0

Table 2.2 Antiviral activity of quercetin, lecithin and quercetin-enriched lecithin against HAV strain (Ramadan & Asker, 2009)

Compound/formulation	Initial virus Count (PFU/mL)	Final virus Count (PFU/mL)	Inhibition (%)
Quercetin (Qu)	0.96×10^7	0.51×10^7	45
Lecithin (Le)		0.63×10^7	33
Qu-Le (3:97, w/w)		0.21×10^7	75

Functional Properties of Other Phenolipids

Lin et al. (2009) prepared a curcumin-encapsulated o/w emulsion using Tween 80, lecithin, and ethyl oleate. The *in vivo* absorption of curcumin after administration exhibited its low bioavailability. The highest level of oil solubilized in the micro-emulsion system was achieved when lecithin/Tween 80 molar ratio was 0.3. The encapsulation of curcumin in emulsion prevented the degradation of curcumin and increased the water phase's curcumin concentration. These functional properties make curcumin suitable for delivery in functional food and nutraceuticals.

Balakrishna et al. (2017) synthesized protected phenolics-PC by the esterification of lyso-PC with protected phenolic acids (Lovell et al., 2012). EDC.HCl was added to protected phenolic acids solutions in CH_2Cl_2, followed by the addition of lyso-PC and DMAP. De-protection of protected phenolic PL with tetra-n-butyl ammonium fluoride was carried out (Li et al., 2014). The prepared phenoylated-PC exhibited showed excellent antioxidant potential, antimicrobial against *Klebsiella planticola*, and high antifungal activity against *Candida albicans*.

Cui et al. (2015) mentioned that amphiphilic PL dioleoylphosphatidylethanolamine (DOPE) and dioleoylphosphatidylcholine (DOPC), could form reverse micelles in oils. They tested how these reverse micelles affect α-tocopherol and Trolox in the stripped and soybean oils. DOPC reverse micelles decreased the activity of Trolox and α-tocopherol, while DOPE increased the antioxidant potential of Trolox and α-tocopherol. Trolox showed better antioxidant potential than α-tocopherol in the presence of DOPE and DOPC micelles due to Trolox partitioned more at the interfaces. Different ratios of DOPC to DOPE were added to oil-containing α-tocopherol, whereas the antioxidant traits increased by increasing the

DOPE/DOPC ratio. The addition of DOPE to soybean oil inhibited lipid oxidation, while DOPC was ineffective. Moreover, HPLC showed that DOPE regenerated α-tocopherol. The investigation concluded that the antioxidant potential of tocopherols might be enhanced using PE to engineer the traits of reverse micelles in oil.

Conclusions

Phenolics, when complexed with PL, give rise to a novel delivery systems. Phytosomes are superior to liposomes due to higher stability and absorption. Phytosomes could play an essential role in efficient drug delivery of a wide spectrum of phytochemicals such as xanthones, flavones, and terpenes. In addition, phytosomes could be developed for several therapeutic applications including anti-inflammatory, cardiovascular, and anticancer traits. Pheno-phospholipids showed novel antimicrobial and antioxidant activities. The stability of quercetin-enriched lecithin was higher than pure lecithin in stabilizing triolein model systems. Triolein samples containing quercetin-lecithin complexes had high antiradical traits. Quercetin-enriched lecithin could be used for novel applications in food and non-food (pharmaceuticals, and cosmetics) industry. Moreover, the antimicrobial and antiviral traits of quercetin-enriched lecithin mixtures were higher thatn pure quercetin or lecithin. Models supplemented with pheno-phospholipids had improved antiradical potential. The synergism of PL and quercetin in the triolein models and the synergy of quercetin, PL, and tocols in oil systems must be considered. Pheno-phospholipids may be used to increase the biological properties, functionally, and health impacts of PL and phenolics in different products.

References

Aditya, N. P., Macedo, A. S., Doktorovova, S., Souto, E. B., Kim, S., Chang, P.-S., et al. (2014). Development and evaluation of lipid nanocarriers for quercetin delivery: A comparative study of solid lipid nanoparticles (SLN), nanostructured lipid carriers (NLC), and lipid nanoemulsions (LNE). *LWT- Food Science and Technology, 59*(1), 115–121. https://doi.org/10.1016/j.lwt.2014.04.058

Althans, D., Schrader, P., & Enders, S. (2014). Solubilisation of quercetin: Comparison of hyperbranched polymer and hydrogel. *Journal of Molecular Liquids, 196*, 86–93. https://doi.org/10.1016/j.molliq.2014.03.028

Anankanbil, S., Pérez, B., Banerjee, C., & Guo, Z. (2018). New phenophospholipids equipped with multi-functionalities: Regiospecific synthesis and characterization. *Journal of Colloid and Interface Science, 523*, 169–178. https://doi.org/10.1016/j.jcis.2018.03.097

Azuma, K., Ippoushi, K., Ito, H., Higashio, H., & Terao, J. (2002). Combination of lipids and emulsifiers enhances the absorption of orally administered quercetin in rats. *Journal of Agricultural and Food Chemistry, 50*, 1706–1712.

Balakrishna, M., Kaki, S. S., Karuna, M. S. L., Sarada, S., Kumar, C. G., & Prasad, R. B. N. (2017). Synthesis and in vitro antioxidant and antimicrobial studies of novel structured phosphatidylcholines with phenolic acids. *Food Chemistry, 221*, 664–672. https://doi.org/10.1016/j.foodchem.2016.11.121

Bandarra, N., Campos, R., Batista, I., Nunes, M. L., & Empis, J. (1999). Antioxidant synergy of α-tocopherol and phospholipids. *Journal of the American Oil Chemists' Society, 76*, 905–913.

Barras, A., Mezzetti, A., Richard, A., Lazzaroni, S., Roux, S., Melnyk, P., et al. (2009). Formulation and characterization of polyphenol-loaded lipid nanocapsules. *International Journal of Pharmaceutics, 379*(2), 270–277. https://doi.org/10.1016/j.ijpharm.2009.05.054

Basaga, H., Poli, G., Tekkaya, C., & Aras, I. (1997). Free radical scavenging and antioxidative properties of 'Silibin' complexes on microsomal lipid peroxidation. *Cell Biochemistry and Function, 15*(1),27–33.https://doi.org/10.1002/(sici)1099-0844(199703)15:1<27::aid-cbf714>3.3.co;2-n

Bombardelli, E. (1991). Phytosome: New cosmetic delivery system. *Bollettino Chimico Farmaceutico, 130*(11), 431–438.

Bombardelli, E. (1994). Phytosomes in functional cosmetics. *Fitoterapia, 65*(5), 320–327.

Bombardelli, E., Curri, S. B., Loggia Della, R., Del, N. P., Tubaro, A., & Gariboldi, P. (1989). Complexes between phospholipids and vegetal derivatives of biological interest. *Fitoterapia, 60*, 1–9.

Bombardelli, E., & Patri, G. F. (1991). Complex compounds of bioflavonoids with phospholipids, their prepcal and cosmetic compositions containing them. U.S. Patent Number 5,043,323.

Bombardelli, E., & Spelta, M. (1991). Phospholipid-polyphenol complexes: A new concept in skin care ingredients. *Cosmetics & Toiletries, 106*(3), 69–76.

Bombardelli, E., Spelta, M., Loggia Della, R., Sosa, S., & Tubaro, A. (1991). Aging skin: Protective effect of silymarin-PHYTOSOME. *Fitoterapia, 62*(2), 115–122.

Bouarab, L., Maherani, B., Kheirolomoom, A., Hasan, M., Aliakbarian, B., Linder, M., et al. (2014). Influence of lecithin-lipid composition on physico-chemical properties of nanoliposomes loaded with a hydrophobic molecule. *Colloids and Surfaces. B, Biointerfaces, 115*, 197–204. https://doi.org/10.1016/j.colsurfb.2013.11.034

Brown, J. E., Khodr, H., Hider, R. C., & Rice-Evans, C. A. (1998). Structural dependence of flavonoid interactions with Cu2+ ions: Implications for their antioxidant properties. *Biochemistry Journal, 330*, 1173–1178.

Calabro, M. L., Tommasini, S., Donato, P., Raneri, D., Stancanelli, R., Ficarra, P., et al. (2004). Effects of alpha- and beta-cyclodextrin complexation on the physicochemical properties and antioxidant activity of some 3-hydroxyflavones. *Journal of Pharmaceutical and Biomedical Analysis, 35*(2), 365–377. https://doi.org/10.1016/j.jpba.2003.12.005

Chen, B., Han, A., Laguerre, M., McClements, D. J., & Decker, E. A. (2011). Role of reverse micelles on lipid oxidation in bulk oils: Impact of phospholipids on antioxidant activity of α-tocopherol and Trolox. *Food & Function, 2*(6), 302–309. https://doi.org/10.1039/c1fo10046g

Citernesi, U., & Sciacchitano, M. (1995). Phospholipids/active ingredient complexes. *Cosmetics & Toiletries, 110*(11), 57–68.

Crauste, C., Vigor, C., Brabet, P., Picq, M., Lagarde, M., Hamel, C., et al. (2014). Synthesis and evaluation of polyunsaturated fatty acid-phenol conjugates as anti-carbonyl-stress lipophenols. *European Journal of Organic Chemistry, 2014*, 4548–4561.

Cui, L., & Decker, E. A. (2016). Phospholipids in foods: Prooxidants or antioxidants? *Journal of the Science of Food and Agriculture, 96*(1), 18–31. https://doi.org/10.1002/jsfa.7320

Cui, L., Mcclements, D. J., & Decker, E. A. (2015). Impact of phosphatidylethanolamine on the antioxidant activity of α-tocopherol and trolox in bulk oil. *Journal of Agricultural and Food Chemistry, 63*, 3288–3294. https://doi.org/10.1021/acs.jafc.5b00243

Damodaran, S., & Parkin, K. L. (2008). *Fennema's food chemistry*. Boca Raton, FL: CRC Press.

Das, S., Mandal, A., Ghosh, A., Panda, S., Das, N., & Sarkar, S. (2008). Nanoparticulated quercetin in combating age related cerebral oxidative injury. *Current Aging Science, 1*, 169–174.

De Paz, E., Martín, A., & Cocero, M. J. (2012). Formulation of β-carotene with soybean lecithin by PGSS (Particles from Gas Saturated Solutions)-drying. *The Journal of Supercritical Fluids, 72*, 125–133. https://doi.org/10.1016/j.supflu.2012.08.007

De Paz, E., Martín, A., Estrella, A., Rodríguez-Rojo, S., Matias, A. A., Duarte, C. M. M., et al. (2012). Formulation of β-carotene by precipitation from pressurized ethyl acetate-on-water emulsions for application as natural colorant. *Food Hydrocolloids, 26*(1), 17–27. https://doi.org/10.1016/j.foodhyd.2011.02.031

Doert, M., Jaworska, K., Moersel, J.-T., & Kroh, L. (2012). Synergistic effect of lecithins for tocopherols: Lecithin-based regeneration of α-tocopherol. *European Food Research and Technology, 235*, 915–928.

Doert, M., Krüger, S., Morlock, G. E., & Kroh, L. W. (2017). Synergistic effect of lecithins for tocopherols: Formation and antioxidant effect of the phosphatidylethanolamine-l-ascorbic acid condensate. *European Food Research and Technology, 243*(4), 583–596. https://doi.org/10.1007/s00217-016-2768-z

Falasca, K., Ucciferri, C., Mancino, P., Vitacolonna, E., De Tullio, D., Pizzigallo, E., et al. (2008). Treatment with silybin-vitamin E-phospholipid complex in patients with hepatitis C infection. *Journal of Medical Virology, 80*, 1900–1906. https://doi.org/10.1002/jmv.21292

Filburn, C. R., Kettenacker, R., & Griffin, D. W. (2007). Bioavailability of a silybin-phosphatidylcholine complex in dogs. *Journal of Veterinary Pharmacology and Therapeutics, 30*, 132–138.

Fraile, M., Buratto, R., Gomez, B., Martín, A., & Cocero, M. J. (2014). Enhanced delivery of quercetin by encapsulation in poloxamers by supercritical antisolvent process. *Industrial & Engineering Chemistry Research, 53*(11), 4318–4327. https://doi.org/10.1021/ie5001136

Giacomelli, S., Gallo, D., Apollonio, P., Ferlini, C., Distefano, M., Morazzoni, P., et al. (2002). Silybin and its bioavailable phospholipid complex (IdB 1016) potentiate in vitro and in vivo the activity of cisplatin. *Life Sciences, 70*(12), 1447–1459. https://doi.org/10.1016/S0024-3205(01)01511-9

Gonçalves, V. S. S., Rodríguez-Rojo, S., De Paz, E., Mato, C., Martín, T., & Cocero, M. J. (2015). Production of water soluble quercetin formulations by pressurized ethyl acetate-in-water emulsion technique using natural origin surfactants. *Food Hydrocolloids, 51*, 295–304. https://doi.org/10.1016/j.foodhyd.2015.05.006.

Gugler, R., Leschik, M., & Dengler, H. J. (1975). Disposition of quercetin in man after single oral and intravenous doses. *European Journal of Clinical Pharmacology, 9*, 229–234.

Hassanien, M. F. R. (2016). Formulation and functionality of phenolipids for novel foods and pharmaceuticals. US20160175377, WO2016097779.

Hoeller, S., Sperger, A., & Valenta, C. (2009). Lecithin based nanoemulsions: A comparative study of the influence of non-ionic surfactants and the cationic phytosphingosine on physicochemical behaviour and skin permeation. *International Journal of Pharmaceutics, 370*(1e2), 181–186. https://doi.org/10.1016/j.ijpharm.2008.11.014

Husain, S. R., Terao, J., & Matsushita, S. (1984). Comparison of phosphatidylcholine and phosphatidylethanolamine affecting the browning reaction of heated oil. In M. Fugimaki, M. Namiki, & H. Kato (Eds.), *Amino-carbonyl reactions in food and biological systems* (pp. 301–309). New York: Elsevier.

Jewell, N. E., & Nawar, W. W. (1980). Thermal oxidation of phospholipids.1,2-Dipalmitoyl-sn-glycerol-3-phosphoethanolamine. *Journal of the American Oil Chemists Society, 57*, 398–402.

Jiang, Y. N., Yu, Z. P., Yan, Z. M., & Chen, J. M. (2001). Studies on preparation of herba epimedii flavanoid phytosomes and their pharmaceutics. *Zhongguo Zhong Yao Za Zhi, 26*, 105–108.

Joshi, A., Paratkar, S. G., & Thorat, B. N. (2006). Modification of lecithin by physical, chemical and enzymatic methods. *European Journal of Lipid Science and Technology, 108*, 363–373.

Judde, A., Villeneuve, P., Rossignol-Castera, A., & le Guillou, A. (2003). Antioxidant effect of soy lecithins on vegetable oil stability and their synergism with tocopherols. *Journal of the American Oil Chemists Society, 80*, 1209–1215.

Khaled, K. A., El-Sayed, Y. M., & Al-Hadiya, B. M. (2003). Disposition of the flavonoid quercetin in rats after single intravenous and oral doses. *Drug Development and Industrial Pharmacy, 29*, 397–403.

Kidd, P., & Head, K. (2005). A review of the bioavailability and clinical efficacy of milk thistle Phytosome: A silybin-phosphatidylcholine complex. *Alternative Medicine Review, 10*(3), 193–203.

Kiełbowicz, G., Gładkowski, W., Chojnacka, A., & Wawrzenczyk, C. (2012). A simple method for positional analysis of phosphatidylcholine. *Food Chemistry, 135*, 2542–2548.

Kim, M. R., Shim, J. Y., Park, K. H., Imm, B. J., Oh, S., & Imm, J. Y. (2009). Optimization of the enzymatic modification of egg yolk by phospholipase A2 to improve its functionality for mayonnaise production. *Food Science and Technology, 42*, 250–255.

Kim, Y., Lee, D., Jung, E., Bae, J., Lee, S., Pyo, H., et al. (2015). Preparation and characterization of quercetin-loaded silica microspheres stabilized by combined multiple emulsion and sol-gel processes. *Chemical Industry and Chemical Engineering Quarterly, 21*(1e1), 85–94. https:// doi.org/10.2298/CICEQ131002010K

King, M. F., Boyd, L. C., & Sheldon, B. W. (1992). Antioxidant properties of individual phospholipids in a salmon oil model system. *Journal of the American Oil Chemists Society, 69*, 545–551.

Kroll, J., Rawel, H., & Rohn, S. (2003). A review. Reactions of plant phenolics with food proteins and enzymes under special consideration of covalent bonds. *Food Science and Technology, 9*, 205–218.

Kumari, A., Yadav, S. K., Pakade, Y. B., Singh, B., & Yadav, S. C. (2010). Development of biodegradable nanoparticles for delivery of quercetin. *Colloids and Surfaces. B, Biointerfaces, 80*(2), 184–192. https://doi.org/10.1016/j.colsurfb.2010.06.002

Laguerre, M., Lecomte, J., & Villeneuve, P. (2015). The use and effectiveness of antioxidants in lipids preservation: Beyond the polar paradox. *Handbook of Antioxidants for Food Preservation, 2*, 349–372. https://doi.org/10.1016/B978-1-78242-089-7.00014-2

Lévai, G., Martín, Á., De Paz, E., Rodríguez-Rojo, S., & Cocero, M. J. (2015). Production of stabilized quercetin aqueous suspensions by supercritical fluid extraction of emulsions. *Journal of Supercritical Fluids, 100*, 34–45. https://doi.org/10.1016/j.supflu.2015.02.019

Li, H. L., Zhao, X. B., Ma, Y. K., Zhai, G. X., Li, L. B., & Lou, H. X. (2009). Enhancement of gastrointestinal absorption of quercetin by solid lipid nanoparticles. *Journal of Controlled Release, 133*, 238–244.

Li, S. N., Fang, L. L., Zhong, J. C., Shen, J. J., Xu, H., Yang, Y. Q., et al. (2014). Catalytic asymmetric synthesis of the Colorado potato beetle pheromone and its enantiomer. *Tetrahedron: Asymmetry, 25*, 591–595.

Lin, C. C., Lin, H. Y., Chen, H. C., Yu, M. W., & Lee, M. H. (2009). Stability and characterisation of phospholipid-based curcumin-encapsulated microemulsions. *Food Chemistry, 116*(4), 923–928. https://doi.org/10.1016/j.foodchem.2009.03.052

Liu, B., Du, J., Zeng, J., Chen, C., & Niu, S. (2009). Characterization and antioxidant activity of dihydromyricetin-lecithin complex. *European Food Research and Technology, 230*(2), 325–331. https://doi.org/10.1007/s00217-009-1175-0

Lovell, J. F., Jin, C. S., Huynh, E., MacDonald, T. D., Cao, W., & Zheng, G. (2012). Enzymatic Regioselection for the synthesis and biodegradation of porphysome nanovesicles. *Angewandte Chemie International Edition, 51*, 2429–2433.

Lucas, R., Comelles, F., Alcantara, D., Maldonado, O. S., Curcuroze, M., Parra, J. L., et al. (2010). Surface-active properties of lipophilic antioxidants tyrosol and hydroxytyrosol fatty Acid esters: A potential explanation for the nonlinear hypothesis of the antioxidant activity in oil-in-water emulsions. *Journal of Agricultural and Food Chemistry, 58*, 8021–8026.

Maiti, K., Mukherjee, K., Gantait, A., Ahamed, H. N., Saha, B. P., & Mukherjee, P. K. (2005). Enhanced therapeutic benefit of quercetin-phospholipid complex in carbon tetrachloride-induced acute liver injury in rats: A comparative study. *Iranian Journal of Pharmacology & Therapeutics Research IJPT, 405*(4), 1735–265742.

Maiti, K., Mukherjee, K., Gantait, A., Saha, B. P., & Mukherjee, P. K. (2006). Enhanced therapeutic potential of naringenin-phospholipid complex in rats. *The Journal of Pharmacy and Pharmacology, 58*(9), 1227–1233.

Makris, D. P., & Rossiter, J. T. (2001). Comparison of quercetin and a non-orthohydroxy flavonol as antioxidants by comparing in vitro oxidation reactions. *Journal of Agricultural and Food Chemistry, 49*, 3370–3377.

Manach, C., Scalbert, A., & Morand, C. (2004). Polyphenols: Food sources and bioavailability. *The American Journal of Clinical Nutrition, 79*, 727–747.

Meng, X., Maliakal, P., Hong, L., Lee, M. J., & Yang, C. S. (2004). Urinary and plasma levels of resveratrol and quercetin in humans, mice, and rats after ingestion of pure compounds and grape juice. *Journal of Agricultural and Food Chemistry, 52*, 935–942.

Mignet, N., Seguin, J., & Chabot, G. G. (2013). Bioavailability of polyphenol liposomes: A challenge ahead. *Pharmaceutics, 5*(3), 457–471. https://doi.org/10.3390/pharmaceutics5030457

Mukherjee, P. K. (2001). Evaluation of Indian traditional medicine. *Drug Information Journal, 35*(2), 623–631.

Mukherjee, P. K. (2002). Problems and prospects for the GMP in herbal drugs in Indian systems of medicine. *Drug Information Journal, 63*(3), 6635–6644.

Mulholland, P. J., Ferry, D. R., Anderson, D., Hussain, S. A., Young, A. M., Cook, J. E., et al. (2001). Pre-clinical and clinical study of QC12, a water-soluble, pro-drug of quercetin. *Annals of Oncology, 12*, 245–248.

Murray, C. W., Booth, A. N., Deeds, F., & Jones, F. T. (1954). Absorption and metabolism of rutin and quercetin in the rabbit. *Journal of the American Pharmaceutical Association, 43*, 361–364.

Naik, S. R., & Panda, V. S. (2007). Antioxidant and hepatoprotective effects of Ginkgo biloba phytosomes in carbon tetrachloride-induced liver injury in rodents. *Liver International, 27*, 393–399. https://doi.org/10.1111/j.1478-3231.2007.01463.x

Nasibullin, R. S., Nikitina, T. I., Afanas'eva, Y., Nasibullin, T. R., & Spirikhin, L. V. (2002). Complex of 3,5,7,30,40-pentahydroxyplavonol with phospatidylcholine. *Pharmaceutical Chemistry Journal, 36*(9), 492–495. https://doi.org/10.1023/A:1021848806762

Nemeth, K., & Piskula, M. K. (2007). Food content, processing, absorption and metabolism of onion flavonoids. *Critical Reviews in Food Science and Nutrition, 47*, 397–409.

Okulus, M., & Gliszczyńska, A. (2020). Enzymatic synthesis of o-methylated phenophospholipids by lipase-catalyzed acidolysis of egg-yolk phosphatidylcholine with anisic and veratric acids. *Catalysts, 10*(5). https://doi.org/10.3390/catal10050538

Parmara, A., Singhb, K., Bahadurc, A., Marangonib, G., & Bahadura, P. (2011). Interaction and solubilization of some phenolic antioxidants in Pluronic® micelles. *Colloids and Surfaces B: Biointerfaces, 86*, 319–326.

Passi, S., Ricci, R., Cataudella, S., Ferrante, I., de Simone, F., & Rastrelli, L. (2004). Fatty acid pattern, oxidation product development, and antioxidant loss in muscle tissue of rainbow trout and Dicentrarchus labrax during growth. *Journal of Agricultural and Food Chemistry, 52*, 2587–2592.

Porter, W. L. (1980). Recent trends in food applications of antioxidants. In M. G. Simic & M. Karel (Eds.), *Autoxidation in food and biological systems* (pp. 295–365). New York: Plenum Press.

Pralhad, T., & Rajendrakumar, K. (2004). Study of freeze-dried quercetin-cyclodextrin binary systems by DSC, FT-IR, X-ray diffraction and SEM analysis. *Journal of Pharmaceutical and Biomedical Analysis, 34*(2), 333–339. https://doi.org/10.1016/S0731-7085(03)00529-6

Priprem, A., Watanatorn, J., Sutthiparinyanont, S., Phachonpai, W., & Muchimapura, S. (2008). Anxiety and cognitive effects of quercetin liposomes in rats. *Nanomedicine, 4*(1), 70–78. https://doi.org/10.1016/j.nano.2007.12.001

Provinciali, M., Papalini, F., Orlando, F., Pierpaoli, S., Donnini, A., Morazzoni, P., et al. (2007). Effect of the silybin-phosphatidylcholine complex (IdB 1016) on the development of mammary tumors in HER-2/neu transgenic mice. *Cancer Research, 67*(5), 2022–2029. https://doi.org/10.1158/0008-5472.CAN-06-2601

Ramadan, M. F. (2008). Quercetin increases antioxidant activity of soy lecithin in a triolein model system. *LWT- Food Science and Technology, 41*(4), 581–587. https://doi.org/10.1016/j.lwt.2007.05.008

Ramadan, M. F. (2012). Antioxidant characteristics of phenolipids (quercetin-enriched lecithin) in lipid matrices. *Industrial Crops and Products, 36*(1), 363–369. https://doi.org/10.1016/j.indcrop.2011.10.008

Ramadan, M. F. (2013). Phenolipids: Novel quercetin-enriched lecithin for functional foods and nutraceuticals. *Inform, 24*(8), 532–537.

Ramadan, M. F. (2014). Phenolipids: New generation of antioxidants with higher bioavailability. *Austin Journal of Nutrition and Food Sciences, 1*(1), 2–3.

Ramadan, M. F., & Asker, M. M. S. (2009). Antimicrobical and antivirial impact of novel quercetin-enriched lecithin. *Journal of Food Biochemistry, 33*, 557–571.

Reddy, K. K., Shanker, K. S., Ravinder, T., Prasad, R. B. N., & Kanjilal, S. (2010). Chemo-enzymatic synthesis and evaluation of novel structured phenolic lipids as potential lipophilic antioxidants. *European Journal of Lipid Science and Technology, 112*, 600–608.

Rice-Evans, C., & Miller, N. J. (1996). Antioxidant activities of flavonoids as bioactive components of food. *Biochemical Society Transactions, 24*, 790–794.

Riemenschneider, W. (2000). Esters, Organic. In Ullmann's Encyclopedia of Industrial Chemistry, (Ed.). https://doi.org/10.1002/14356007.a09_565

Rombaut, R., & Dewettinck, K. (2006). Properties, analysis and purification of milk polar lipids. *International Dairy Journal, 16*, 1362–1373.

Russo, M., Spagnuolo, C., Tedesco, I., Bilotto, S., & Russo, G. L. (2012). The flavonoid quercetin in disease prevention and therapy: Facts and fancies. *Biochemical Pharmacology, 83*(1), 6–15. https://doi.org/10.1016/j.bcp.2011.08.010

Selmair, P. L., & Koehle, P. (2009). Molecular structure and baking performance of individual glycolipid classes from lecithins. *Journal of Agriculture and Food Chemistry, 57*, 5597–5609.

Semalty, A., Semalty, M., & Rawat, M. S. M. (2007). The phyto-phospholipid complexes-phytosomes: A potential therapeutic approach for herbal hepatoprotective drug delivery. *Pharmacognosy Reviews, 1*(2), 369–374.

Shahidi, F., & Zhong, Y. (2011). Revisiting the polar paradox theory: A critical overview. *Journal of Agricultural and Food Chemistry, 59*(8), 3499–3504. https://doi.org/10.1021/jf104750

Somsuta, C., Sami, S., Nirmalendu, D., Somsubhra, T. C., Swarpupa, G., & Snehasitka, S. (2012). The use of nano-quercetin to arrest mitochondrial damage and MMP-9 upregulation during prevention of gastric inflammation induced by ethanol in rat. *Biomaterials, 33*, 2991–3001.

Souza, M. P., Vaz, A. F. M., Correia, M. T. S., Cerqueira, M. A., Vicente, A. A., & Carneiro-da-Cunha, M. G. (2013). Quercetin-loaded lecithin/chitosan nanoparticles for functional food applications. *Food and Bioprocess Technology, 7*(4), 1149–1159. https://doi.org/10.1007/s11947-013-1160-2

Srinivas, K., King, J. W., Howard, L. R., & Monrad, J. K. (2010). Solubility and solution thermo-dynamic properties of quercetin and quercetin dihydrate in subcritical water. *Journal of Food Engineering, 100*(2), 208–218. https://doi.org/10.1016/j.jfoodeng.2010.04.001

Terao, J., & Piskula, M. K. (1999). Flavonoids and membrane lipid peroxidatiotn inhibition. *Nutrition, 15*, 790–791.

van Acker, S. A. B. E., van de Berg, D.-J., Tromp, M. N. J. L., Griffioen, D. H., van Bennekom, W., van der Vijgh, W. J. F., et al. (1996). Structural aspects of antioxidant activity of flavonoids. *Free Radical Biology and Medicine, 20*(3), 331–342.

Van Nieuwenhuyzen, W., & Tomás, M. C. (2008). Update on vegetable lecithin and phospholipid technologies. *European Journal of Lipid Science and Technology, 110*, 472–486.

Varona, S., Martín, A., & Cocero, M. J. (2009). Formulation of a natural biocide based on lavandin essential oil by emulsification using modified starches. *Chemical Engineering and Processing: Process Intensification, 48*(6), 1121–1128. https://doi.org/10.1016/j.cep.2009.03.002

Vikbjerg, A. F., Rusig, J. Y., Jonsson, G., Mu, H., & Xu, X. (2006). Comparative evaluation of the emulsifying properties of phosphatidylcholine after enzymatic acyl modification. *Journal of Agricultural and Food Chemistry, 54*(9), 3310–3316. https://doi.org/10.1021/jf052665w

Vitaglione, P., Morisco, F., Caporaso, N., & Fogliano, V. (2005). Dietary Antioxidant Compounds and Liver Health. *Critical Reviews in Food Science and Nutrition, 44*, 575–586.

Wu, T.-H., Yen, F.-L., Lin, L.-T., Tsai, T.-R., Lin, C.-C., & Cham, T.-M. (2008). Preparation, physicochemical characterization, and antioxidant effects of quercetin nanoparticles. *International Journal of Pharmaceutics, 346*(1e2), 160–168. https://doi.org/10.1016/j.ijpharm.2007.06.036

Yang, H., Mu, Y., Chen, H., Xiu, Z., & Yang, T. (2013). Enzymatic synthesis of feruloylated lysophospholipid in a selected organic solvent medium. *Food Chemistry, 141*, 3317–3322.

Yanyu, X., Yunmei, S., Zhipeng, C., & Quineng, P. (2006). The preparation of silybinphospholipid complex and the study on its pharmacokinetics in rats. *International Journal of Pharmaceutics, 307*(1), 77–82.

Yuan, Z. P., Chen, L. J., Fan, L. Y., Tang, M. H., Yang, G. L., Yang, H. S., et al. (2006). Liposomal quercetin efficiently suppresses growth of solid tumors in murine models. *Clinical Cancer Research, 12*, 3193–3199.

Zheng, Y., & Chow, A. H. L. (2009). Production and characterization of a spray-dried hydroxypropyl-beta-cyclodextrin/quercetin complex. *Drug Development and Industrial Pharmacy, 35*(6), 727–734. https://doi.org/10.1080/03639040802526805

Chapter 3
Chemistry and Functionality
of Lipo-phenolics

Abstract Natural phenolics are novel bioactive compounds, but their biological (antioxidant, antiradical, and antimicrobial) properties are limited due to their hydrophilic character. The lipophilicity of phenolics could be modified by attaching a lipophilic moieties to the phenolic compounds (lipo-phenolics) to change its hydrophilic/lipophilic balance. Lipo-phenolics already proved to be excellent antioxidants in food and cosmetics. Lipophilization allows the formulation of new functionalized bioactives having useful traits compared to natural hydrophilic phenolics. The lipophilization of phenolics enhanced their solubility in apolar media. Lipophilization changes the location of the new compound in the emulsions, which might increase their antioxidant activities. Therefore, in the emulsified systems, lipo-phenolics are suggested to locate at the lipid/water phase interface and to increase the protection of fats and oils. Lipophilization reaction could be obtained chemically or enzymatically. This chapter reported on the chemistry, preparation, and functionality of lipid-enriched phenolics (lipo-phenolics), broadening their food, pharmaceuticals, and cosmetics applications. The strategies of the lipophilization of phenolics, the impact of modification on the biological traits, and potential uses of the resulting lipo-phenolics are reviewed.

Keywords Phenolipids · Amphiphiles · Phenolic lipids · Lipophilization · Antioxidant activity · Esterification · Hydrophobicity · Lipophilicity · Lipophilic antioxidants

Abbreviations

ABTS	2,2′-azinobis(3-ethylbenzothiazoline-6-sulphonic acid)
BHT	Butylated hydroxytoluene
CA	Caffeic acid
CLA	Conjugated linoleic acid
EGCG	Epigallocatechin gallate
FA	Ferulic acid
FDB	1(3)-feruloyl-dibutyryl-glycerol
FFA	Free fatty acids

© The Author(s), under exclusive license to Springer Nature
Switzerland AG 2021
M. F. Ramadan, *Pheno-phospholipids and Lipo-phenolics*,
https://doi.org/10.1007/978-3-030-67399-4_3

FMB   1(3)-feruloyl-monobutyryl-glycerol
PUFA   Polyunsaturated fatty acids
R   Resveratrol
RA   Rosmarinic acid
RDAG   1,2-diacylglycerol rosmarinate
ROS   Reactive oxygen species
THF   Tetrahydrofuran

Introduction

Phenolics and phenol derivatives are found in nature, especially in the plant king-dom. Phenolic acids, flavonoids, stilbenes, and lignans are secondary plant metabo-lites with biological and functional properties. Phenolic compounds are essential antioxidants and play an essential role in the biological systems (Durand et al., 2015). Antioxidants prevent oxidation in food, cosmetics, and pharmaceuticals. In the edible applications, the requirements to select active antioxidant are to under-stand the protective traits of the antioxidants, the location of the antioxidant in the system, to identify the oxidized substrates, and the impacts of other components on the antioxidant potential (Figueroa-Espinoza & Villeneuve, 2005; Frankel & Meyer, 2000). The antioxidants with surface-active characteristics possess a higher ability to delay oxidation in the emulsion systems because it might concentrate at the oil-water interface (Lucas et al., 2010). Stasiuk and Kozubek (2010) reviewed the natu-ral phenolic-lipids derived from mono and dihydroxy phenols. Due to their high amphiphilic properties, phenolic-lipids could incorporate into liposomal mem-branes and erythrocytes. Phenolic-lipids' ability to delay or inhibit fungal, bacterial, parasite and protozoan growth depend upon their interaction with protein and their membrane-disturbing traits.

The polar paradox theory mentioned that lipophilic antioxidants are more active in the oil-in-water (o/w) emulsion than other hydrophilic antioxidants. In contrast, hydrophilic antioxidants are more active in bulk oils (Alemán et al., 2015). Based on this theory, phenolics act better in the bulk oils than in the emulsions. The lipo-philization of phenolics with alkyl chain with different lengths would reduce their polarity and therefore change their distribution between the emulsion phases. Thus, lipophilization of phenolics is anticipated to enhance the antioxidant impact of phe-nolics. The amphiphilic trait of these called "phenolipids" (Laguerre et al., 2010) could envisage surface-active traits that could lead to a non-ionic surfactant. Different polar head groups were utilized in non-ionic surfactants, such as amino acids and carbohydrates, leading to n-alkyl polyglucosides (von Rybinski & Hill, 1998), sorbitan esters (Cottrell & van Peij, 2004), sugar fatty acid esters (Plat & Linhardt, 2001), and amino acid-based surfactant (Moran et al., 2004). Phenolics have been utilized as polar heads of surfactants such as the alkyl esters of p-hydroxyphenyl acetic acid (Yuji et al., 2007) and chlorogenic fatty acid (Lucas et al., 2010; Sasaki et al., 2010).

Because phenolics are polar, lipophilization might extend phenolics' applications in oil-containing food, emulsions, and cosmetics (Laguerre et al., 2010; Ramadan, 2008, 2012). There is an interest in enhancing the amphiphilic traits of phenolics by lipophilization, which corresponds to the grafting of an aliphatic *via* lipase-catalyzed esterification. The operational conditions and strategies can be adapted to perform lipase-mediated hydrophobation (Lopez-Giraldo et al., 2007). Lipophilized phenolics (lipo-phenolics) were developed to enhance their capability to counteract lipid peroxidation (Figueroa-Espinoza & Villeneuve, 2005).

Lipo-phenolics resulting from lipid grafting on a phenolic moieties were formulated through synthesis methods including esterification (Durand et al., 2015; Laguerre et al., 2010; Wang, Hwang, & Lim, 2016; Wang, Zhang, Zhong, Perera, & Shahidi, 2016). A multitude of lipo-phenolics has been prepared. Flavonoids, phenolic acids, and tocols have been lipophilized with different alkyl chain lengths (Buisman et al., 1998; Lecomte, Giraldo, Laguerre, Barea, & Villeneuve, 2010; López Giraldo, Laguerre, Lecomte, FigueroaEspinoza, et al., 2007; Priya, Venugopal, & Chadha, 2002; Stamatis, Sereti, & Kolisis, 1999). In the biological systems, lipophilization might contribute to easier penetration of antioxidant *via* cell membrane that could increase the bioavailability. Therefore, lipo-phenolics provide a perspective of novel compounds with improved or maintained functional traits of food, cosmetics, and pharmaceuticals (Durand et al., 2015).

Phenolics as Natural Antioxidants

Many resources of natural antioxidants are commercially available. Besides, phytoextracts are rich in phenolics could be obtained from fruits, herbs, and vegetables. High addition is commonly needed to exhibit high activities, which might be expensive, and they could impart off-flavors or color changes to the final products (Figueroa-Espinoza & Villeneuve, 2005). Phenolic compounds are of essential interest from food preservation and dietary antioxidant supplements (Halliwell, Aeschbach, Loliger, & Aruoma, 1995). Hydroxyl groups linked with phenolics are the most effective radical scavengers. The efficiency of phenolics depend on their structure, heat stability, volatility, and pH sensitivity (Cuvelier, Richard, & Berset, 1992; Decker, 1998; Figueroa-Espinoza & Villeneuve, 2005; Scholz, Heinrich, & Hunkler, 1994).

Lipid are usually exist in food as dispersions, wherein the physical traits of lipids and water or air interfaces affect oxidative stability. The antioxidative characteristics in multiphase food are influenced by several factors determined by phenomena governing the orientation and localization of antioxidant, by partitioning between the water phase and lipophilic phase, and by interacting with the emulsifiers at the interface (Frankel & Meyer, 2000). Another critical factor is the solubility of the antioxidant concerning the oxidation site (Decker, 1998; Figueroa-Espinoza & Villeneuve, 2005).

Types of Lipophilization

Applications for phenolics in oil-rich cosmetics and foodstuffs are limited because of phenolics' low solubility. The hydrophobicity of phenolics could be improved by enzymatic or chemical lipophilization, by esterifying the carboxylic acid function of phenolic acid with a fatty alcohol, to obtain an amphiphilic compound (Ha, Nihei, & Kubo, 2004; Nihei, Nihei, & Kubo, 2004). To increase oil stability, lipo-phenolics might accumulate at oil-air or oil-water interfaces where oxidation occurs. Enzymatic and chemical esterification of phenolics with alcohols or acids has been studied (Figueroa-Espinoza & Villeneuve, 2005; Lecomte et al., 2010). The chemical synthesis is difficult because of thermal sensitivity and oxidation susceptibility in phenolic acid-alkaline media (Chalas et al., 2001; Nakayama, 1995; Nihei et al., 2004; Taniguchi, Nomura, Chikuno, & Minami, 1997). Moreover, chemical reactions are unselective, involve intermediary and purification steps, as well as generating wastes (Hills, 2003). On the other side, the enzymatic synthesis provides mild conditions, minimization of side reactions and formation of byproducts, a wide variety of pure compounds, fewer intermediary stages and purification steps, and an environmental-friendly process (Hills, 2003; Villeneuve, Muderhwa, Graille, & Haas, 2000). In addition, it is acceptable to use lipases in low water media to esterify fatty alcohol (act as a nucleophile) to a phenolic moiety (Guyot et al., 2000; Guyot, Bosquette, Pina, & Graille, 1997; Silva et al., 2000). A device to recycle the reaction mixtures including phenolic acid and fatty alcohol in optimized composition could be adapted to reach the maximum esterification (Figueroa-Espinoza & Villeneuve, 2005).

Phenolic Acid's Lipophilization

Phenolic acids are hydrophilic active compounds found in spices, fruits, vegetables, and herbs. The phenolic acid family contains the benzoic (C6-C1) acid and cinnamic (C6-C3) acid derivatives. Phenolic acids contain a benzene ring substituted with methoxy or hydroxy groups and a carboxylic group. Phenolic acids are of essential interest because of their functional traits, including antiradical, antioxidant, anti-inflammatory, chelating, antimicrobial, antiallergic, antiviral, and anticarcinogenic traits (Figueroa-Espinoza & Villeneuve, 2005). The hydrophilicity of phenolic acid reduces their antioxidative effectiveness. Thus, it is essential to enhance the solubility of phenolic acids to extend its use as antioxidants. Lipophilization of phenolic acid by esterification with an acyl or alkyl donor has been considered to produce amphiphilic antioxidants (Sørensen, Villeneuve, & Jacobsen, 2017; Szydłowska-Czerniak, Rabiej, & Krzemiński, 2018; Szydłowska-Czerniak, Rabiej, Kyselka, Dragoun, & Filip, 2018).

Antioxidative potential of lipo-phenolics depends on the chemical structure, in particular the distribution and number of methoxyl and/or hydroxyl groups in the aromatic ring. Alkyl esters of hydroxycinnamic (ferulates and caffeates) and benzoic (salicylates, gallates, and vanillates) acids are safe and could be used in foodstuffs, pharmaceuticals, and cosmetics (Figueroa-Espinoza, Laguerre, Villeneuve, & Lecomte, 2013). Octyl ferulate and octyl caffeate exhibited anticancer traits; thus these lipo-phenolics could be added to fatty products (Fiuza et al., 2004; Jayaprakasam, Vanisree, Zhang, Dewitt, & Nair, 2006; Szydłowska-Czerniak, Rabiej, & Krzemiński, 2018; Szydłowska-Czerniak, Rabiej, Kyselka, et al., 2018). Besides, lipo-phenolics showed strong antioxidant traits against lipid oxidation in heterogeneous systems. Antiinflammatory, anticancer, antioxidant, and antibacterial traits of lipo-phenolics added to foodstuffs as well as the relationships between their antioxidative effect and hydrophobicity were studied (Alemán et al., 2015; Durand et al., 2017; Durand, Lecomte, & Villeneuve, 2017; Sørensen et al., 2012; Sørensen et al., 2017; Sørensen, Lyneborg, Villeneuve, & Jacobsen, 2015; Sørensen, Nielsen, Yang, Xu, & Jacobsen, 2012; Szydłowska-Czerniak, Rabiej, & Krzemiński, 2018; Szydłowska-Czerniak, Rabiej, Kyselka, et al., 2018; Wang & Shahidi, 2014).

Enzymatic lipophilization is a novel technique to expand the application of phenolics in lipophilic system, such as the enzymatic lipophilization of rowanberry extracts and epigallocatechin gallate (Aladedunye et al., 2015; Luo et al., 2017; Zhong, Ma, & Shahidi, 2012; Zhong & Shahidi, 2011). There are evidences supporting a relationship between lipo-phenolic's alkyl chain length and their interaction with membrane. Lipo-phenolics were prepared chemo-enzymatically by lyophilizing phenolic acids with sterols, wherein the resulting molecules showed high antioxidant potential (Balakrishna et al., 2017; Wang & Cao, 2011; Wang, Hwang, et al., 2016; Wang, Zhang, et al., 2016).

Esters of gallic acid (i.e., propyl, and octyl gallates) are utilized as food antioxidants (Kubo et al. 2001; Kubo, Masuoka, Xiao, & Haraguchi, 2002; Kubo et al. 2001). Octyl gallate exhibits antimicrobial and antifungal properties (Fujita & Kubo, 2002). Patents have been published regarding the synthesis of lipo-phenolics (phenolic acid lipophilic estes). Nakanishi, Oltz, and Grunberger (1989) synthesized esters of caffeic acid (CA) with a butyl, phenethyl, hexyl, or ethyl alcohol exhibiting anticarcinogenic and antiinflammatory traits. Vercauteren et al. (Vercauteren, Weber, Bisson, & Bignon, 1994) reported the chemical esterification of flavanols with fatty acids with antiradical, lipophilic, and antioxidant traits. Taniguchi et al. (1997) prepared alkyl ferulates by reacting 2-ethyl-1-hexanol with ferulic acid (FA) in the presence of toluene sulfonic acid. Cheetham and Banister (2000) prepared hexyl caffeate (as sunscreen agent) from trans-esterification of chlorogenic acid using lipase with hexane-1-ol. Hexyl caffeate, as a sunscreen agent, exhibited high absorbance of UV light. Nkiliza (1998) esterified grape seed phenolic extract using fatty acid chloride, to be utilized in cosmetics or pharmaceuticals (Figueroa-Espinoza & Villeneuve, 2005).

Antioxidant Potential of Lipo-phenolics

Phenolics act as a free radical quenchers due to their hydrogen-donating capability. The stabilization of phenolics by functional groups in the structure improves the antioxidant potential. The addition of *o*- or *p*-hydroxyl group or *ortho* substitution with electron donor (methoxy or alkyl) groups improves the antioxidant traits (Cuvelier et al., 1992; Figueroa-Espinoza & Villeneuve, 2005).

Silva et al. (2000) measured the antiradical potential of CA and dihydrocaffeic acid esters; wherein *n*-alkyl esters had high antiradical traits. Silva, Borges, and Ferreira (2001) investigated the oxidative stability of sunflower oil enriched with propyl hydro-caffeate (PHC), propyl caffeate (PC), propyl isoferulate (PI), R-tocopherol (R-TOH), propyl ferulate (PF), and propyl gallate (PG). The order of antioxidative effectiveness noted by the Rancimat test was PG > PHC > PC ≫ α-TOH > PI > PF. The compounds having higher antioxidant activities (PG, PC, and PHC) showed higher antiradical potential (Figueroa-Espinoza & Villeneuve, 2005). Chalas et al. (2001) investigated the inhibition of copper-catalyzed LDL oxidation, the radical attack of erythrocyte membrane by phenolic ethyl esters. No relationship between lipophilicity and anti-oxidant potential was noted. The sinapate, caffeate, and ferulate ethyl esters were more inhibitors of hydroperoxide formation and apoprotein oxidation than their cor-responding acids. Nenadis, Zhang, and Tsimidou (2003) revealed that ethyl ferulate was more active than FA as an antioxidant in the oil and o/w emulsion. Neudörffer, Bonnefont-Rousselot, Legrand, Fleury, and Largeron (2004) studied the relation-ship between the oxidation effect of monomers of 4-hydroxycinnamic ethyl esters, together with related dehydrodimers, and their antioxidative effect to delay copper-catalyzed oxidation of LDL. The protective impact was decreased in the following order: sinapate > ferulate > *p*-coumarate esters.

Lipophilization expanded the uses of hydrophilic compounds to lipophilic sys-tems (Bernini et al., 2012; Bernini, Mincione, Barontini, & Crisante, 2008; Chen et al., 2017; Laguerre et al., 2008). Zhong et al. (2012) prepared derivatives of lipo-philized epigallocatechin gallate (EGCG) and tested their antioxidative and antiviral properties (Chen et al., 2017). Kubo et al. (2002) reported that dodecyl gallate, a permitted antioxidant in food, exhibited a preventive antioxidant and high chain-breaking activity. Dodecyl gallate was more active than gallic acid in delaying the oxidation of linoleic acid (Aruoma, Murcia, Butler, & Halliwell, 1993). In addition, dodecyl gallate delayed the lipoxygenase linoleic acid peroxidation; it was active in inhibiting the generation of superoxide anion. Kikuzaki, Hisamoto, Hirose, Akiyama, and Taniguchi (2002) studied the antioxidative traits of hydroxycinnamic acid and FA esters, as well as alkyl gallates and gallic acid, in bulk and multiphase systems. Esterification of FA increased the antiradical in oil (Figueroa-Espinoza & Villeneuve, 2005). Ha et al. (2004) found that octyl gallate is an inhibitor of soybean lipoxygenase, and it was more active than tocopherols in delaying lipid oxidation. Anselmi et al. (2004) tested the antiradical properties of FA and alkyl ferulates. The alkyl ferulates acted with a similar mechanism in membrane models, and the side chain was involved in the interaction with the membrane bilayer.

Several reports have mentioned that the polar paradox does not accurately predict the behavior of antioxidant, and thus the polar paradox hypothesis needs to be reinvestigated (Laguerre et al., 2009, 2010; Panya et al., 2012; Sørensen, Nielsen, et al., 2012; Sørensen, Petersen, et al., 2012). Laguerre et al. (2009) tested the antioxidative effect of chlorogenate esters in a tung oil in water emulsions. A non-linear behavior of the antioxidant effect was noted. They noted an increase in the oxidative impact with increasing alkyl chain length to 12 carbon. The same non-linear behavior was noted with lipophilized rosmarinate (Laguerre et al., 2010). Panya et al. (2012) investigated the antioxidant effect of rosmarinate alkyl esters in soybean oil in the water emulsion. The rosmarinates with shorter alkyl chains were much better antioxidants than rosmarinates with longer fatty alkyl chains. Sørensen, Petersen, et al. (2012) investigated the impact of lipophilized rutin and dihydrocaffeic acid in fish oil-supplemented milk. The medium-chain esters showed higher antioxidant effects than the long-chain esters as well as the pure phenolics (Alemán et al., 2015).

Choo and Birch (2009) and Choo, Birch, and Stewart (2009) conducted lipase-catalyzed trans-esterification of triolein and flaxseed oil with cinnamic acid and FA using lipase. Lipophilized FA showed higher radical scavenging activies than lipophilized cinnamic acid. Laguerre et al. (2009, 2010) investigated the antioxidative effect of lipohilized chlorogenic and rosmarinic acids. The antioxidant effect of chlorogenic acid increased as the alkyl chain length increased (Sørensen, Nielsen, et al., 2012). They hypothesized a so-called "cut-off" effect related to the length of the lipid chain linked to phenolic acid. Regarding rosmarinate esters, the octyl rosmarinate enhanced the antioxidant effect 8 times compared to rosmarinic acid (RA). Lipophilization with medium chain fatty acids was a unique route to increase the antioxidative potential (Laguerre et al., 2010; Sørensen, Nielsen, et al., 2012). A study on lipophilized dihydrocaffeic acids and their antioxidant impact in oil in water emulsion investigated by Sørensen, Petersen, et al. (2012) hypothesized that lipophilized dihydrocaffeic acid tended to follow the hypothesized cut-off effect. The cut-off effect might be specific for the some lipo-phenolic, i.e., the optimal chain length might different between phenolics (Sørensen, Nielsen, et al., 2012).

Research on lipo-phenolics has paid attention to their preparation, *in vitro* antioxidant potential, and impact in model systems. In more complicated systems such as food, antioxidant behavior could be affected by interaction with other ingredients (Shahidi & Zhong, 2011; Sørensen et al., 2008; Sørensen, Nielsen, et al., 2012). Partitioning and antioxidative effect of CA and caffeates were affected by the presence of tocols and the emulsifier type. Sørensen et al. (2017) reported different antioxidant-antioxidant and emulsifier-antioxidant interactions that influenced the efficacy of CA and caffeates as antioxidant compounds in emulsion systems. The hypotheses about antioxidant compounds in emulsion systems based on simple emulsions without the presence of tocols (Sørensen et al., 2017). In addition, lipophilic tyrosyl esters with increasing unsaturation levels were formulated (Pande & Akoh, 2016). The results from crude oil and oil-in-water emulsions suggest that the constructed lipo-phenolics might be utilized as antioxidants. The addition of an acyl moiety increased the antioxidant potential of tyrosol, while increasing the unsaturation level in the acyl moiety was not proportional to its antioxidative effect.

The formulation of lipophilic derivatives of FA *via* esterification or trans-esterification of aliphatic molecules was applied to increase its solubility in a hydrophobic environment. Hydrophobic derivatives (i.e., octyl ferulate), have a better antioxidative impact (Fang, Shima, Kadota, Tsuno, & Adachi, 2006). Besides, triterpene alcohol monoesters such as 24-methylenecycloartenyl ferulate and cycloartenyl ferulate displayed high antioxidant activities and inhibited oxidation effectively than FA (Kikuzaki, Hisamoto, Hirose, Akiyama, and Taniguchi, 2002; Zheng et al., 2010). Laguerre et al. (2011) reported that chlorogenic acid could be lipophilized by fatty alcohols to obtain alkyl chlorogenate esters, which constitutes novel phenolic model to test the effect of the hydrophobicity on the antioxidative activity.

Preparation of Lipo-phenolics

Lipid-enriched phenolics (lipo-phenolics) are amphipathic molecules that display functional traits (Figueroa-Espinoza et al., 2013). Lipo-phenolics conjugate the well-established biological characteristics of phenolics (Pereira et al., 2017) with the improved drug-like traits conferred by lipophilic moiety (Lopez-Giraldo et al., 2007; Laguerre, Bayrasy, Lecomte et al. 2013; Laguerre, Bayrasy, Panya et al., 2013; Pereira et al., 2017).

Choo and Birch (2009) and Choo et al. (2009) prepared lipo-phenolics from lipase-catalyzed trans-esterification of flaxseed oil and triolein with cinnamic acid and FA. Triolein was dissolved in *n*-hexane, cinnamic acid, and FA were added, followed by Novozym 435. The mixture was incubated at 50 °C under nitrogen, with continuous shaking for 2 weeks at 120 rpm. The reaction was stopped by termination with immobilized lipase. Sørensen et al. (2015) synthesized ferulates in an acid-catalyzed reaction with FA and fatty alcohols with alcohol in excess as a reaction medium. Szydłowska-Czerniak, Rabiej, and Krzemiński (2018) synthesized phenolic acid esters with 1-octanol, sulphuric acid solution in 1-octanol. The mixtures were stirred and incubated at 100 °C. Octyl caffeate and octyl ferulate were prepared by Fischer esterification between 1-octanol and the phenolic acid.

Lipophilization with Alcohols

The esterification processes of fatty alcohols with phenolic acids was summarized (Table 3.1). Guyot et al. (1997, 2000) formulated phenolic esters by esterification of fatty alcohols using a thermostable lipase B (Table 3.1). Phenolic acids (i.e, cinnamic, ferulic, caffeic, dihydrocaffeic, 3,4-dimethoxycinnamic, and chlorogenic acid) and alcohols (dodecanol, butanol, octanol, and oleyl alcohol) were esterified without solvent. Regarding hydrocaffeic acid, yields were high with all alcohols. Regarding cinnamic acid, the greatest yields were reached with octanol and butanol.

Table 3.1 Conditions of enzymatic lipophilization of phenolic acids with alcohols

Conditions	Phenolic acid	Fatty alcohol	References
Organic solvent Tannase 40 °C 12 h	Gallic	1-propanol	Yu, Li, and Wu (2004)
Lipase, esterase, or cutinase 15 °C 12 days	Ferulic, cinnamic, p-coumaric, p-hydroxyphenylpropionic	1-octanol	Stamatis, Sereti, and Kolisis (2001)
CTAB/hexane/ pentanol w/o microemulsion Feruloyl esterase 40 °C pH 6.0 8 h	Ferulic	n-pentanol	Giuliani et al. (2001)
Solvent: 2-methyl-2-propanol Lipase 60 °C, under nitrogen 13 days	Ferulic	1-octanol	Compton, Laszlo, and Berhow (2000)
Solvent: 2-methyl-2-propanol Lipase (1.2% w/w) 60 °C 30 days	Chlorogenic	Octanol, dodecanol, hexadecanol, 9-octadecen-1-ol	Guyot et al. (2000)
Lipase (1.5% w/w) 60 °C 30 days	Chlorogenic	Octanol, dodecanol, hexadecanol, 9-octadecen-1-ol	Guyot et al. (2000)
Lipase (150 mg) 50 °C 12 days	Cinnamic, caffeic, o-coumaric, m-coumaric, p-coumaric, ferulic, p-hydroxybenzoic	1-octanol with *C. Antarctica* or *R. miehei*	Stamatis et al. (1999)
Solvent: Diethyl ether, *tert*-butyl methyl ether, 1-butanol, cyclohexane, n-pentane Lipase 34 or 50 °C 5 days	Cinnamic	1-butanol: Diethyl ether, *tert*-butyl methyl ether, 1-butano, cyclohexane, n-pentane, 1-hexanol: n-pentane, 1-dodecanol: n-pentane	Buisman et al. (1998)
Solvent: Diethyl ether Lipase 34 °C >7 days	p-Hydroxybenzoic, gentisic, gallic, vanillic, syringic	1-hexanol	Buisman et al. (1998)
Lipase (2.5%, w/w) 60 °C 15 days	Caffeic, cinnamic, dihydrocaffeic, 3,4-dimethoxycinnamic, ferulic	n-butanol	Guyot et al. (1997)

Adapted from Figueroa-Espinoza and Villeneuve (2005)

More chlorogenic acid was esterified with octanol and hexadecanol than with dodecanol (Guyot et al., 1997, 2000). Esterification was achieved in the cinnamic series when the side chain was saturated and when the aromatic cycle was not para-hydroxylated (Buisman et al., 1998; Stamatis et al., 2001). Stamatis et al. (1999) studied lipase esterification of cinnamic acid derivatives and ascorbic acid with

1-octanol in different solvents. They observed high yields recovered in the solvent-free system. Cinnamic acid esterification yield was better for lipase when applying long-chain alcohol (Figueroa-Espinoza & Villeneuve, 2005).

Buisman et al. (1998) used lipases to esterify benzoic and cinnamic acid derivatives with alcohols (i.e., C4, C6, C12) to formulate lipo-phenolics in solvents (Table 3.1). They reported low ester recovery when utilizing lipases from *C. cylindracae, H. lanuginosa, Rhizopus, C. rugosa, G. candidum,* and *Pseudomonas* species. Using of a low-polar solvent (i.e., *n*-pentane) increased the esterification. The best production was achieved by by performing the reaction in 1-butanol (Guyot et al., 1997), by performing the reaction in 1-butanol. The solvent-free system is preferable due to the cost of eliminating solvent traces. The authors confirmed the inhibiting impact of the electron-donating compounds conjugated to the carboxylic group in the phenolic acid (Figueroa-Espinoza & Villeneuve, 2005).

Stamatis et al. (1999, 2001), Compton et al. (2000), Giuliani et al. (2001), and Topakas et al. (2003) reported on FA lipophilization. Stamatis et al. (1999) esterified, using lipases, FA with 1-octanol. Compton et al. (2000) used lipase from *C. antarctica* to prepare *n*-octyl ferulate at 60 °C in 2-methyl-2-propanol (Table 3.1). Giuliani et al. (2001) utilized the feruloyl esterase to synthesize pentylferulate (Table 3.1). Topakas et al. (2003) studied the synthetic activities of a feruloyl esterase on the esterification of phenolic acids with 1-propanol (Table 3.1). The system was stable and served as an appropriate medium for enzymatic catalyses (Khmelnitsky, Hilhorst, & Veeger, 1988). The enzyme showed no activity on cinnamic acid, showing the importance of a hydroxyl group on the C-4 phenol ring.

In a free-solvent mixture, Guyot et al. (1997) studied lipophilization of FA using *C. antarctica* lipase, and Stamatis et al. (1999, 2001) reproted that lipase from *R. miehei* is an excellent option to recover a 30% esterification yield of FA. Likely, the low solubility of FA in non-polar solvents such as the fatty alcohols affected the enzymatic lipophilization (Guyot et al., 1997), and using more polar solvent is important to facilitate the reaction (Compton et al., 2000). Using of a FA specific esterase solubilized in a w/o micro-emulsion is the good way for bioconversion involving a hydrophobic product and a hydrophilic substrate (Giuliani et al., 2001).

Yu et al. (2004) formulated esters of alcohols and gallic acid in solvents using a chitosan-alginate encapsulated tannase (Table 3.1). The encapsulated tannase was more active than the free enzyme. They studied the effects of the alcohol chain length on the production of gallic acid esters. The optimum production was achieved with alcohols ranging from C3 to C5. The intrinsic enzyme selectivity, the solubility of substrates and products, and the enzyme stability were correlated to the added solvents (Figueroa-Espinoza & Villeneuve, 2005). Immobilized enzymes were better option to lipophilize phenolic acids, because of the facility of recovering the product and the enzyme separately.

Preparation of Lipophilized Oils

Luo et al. (2017) prepared lyophilized oil using blueberry anthocyanin extract dissolved in polyxoyethylene stearate. Novozym 435, 4A molecular sieves, and camellia seed oil were added to the anthocyanin solution then placed at 50 °C. Water was added to remove the redundant anthocyanin, stirred mildly, then centrifuged, and the water layer was discarded. Chen et al. (2017) enzymatically lipophilized epicatechin with FFA from camellia seed oil. To prepare lipophilized camellia oil, the oil was placed in a brown vessel. Epicatechin solution (epicatechin dissolved in polyxoyethylene stearate), molecular sieves and Novozym 435 were added, and the vessel was placed at 50 °C in a shaker. Chemical bonding between phenolics and n-3 polyunsaturated fatty acids (PUFA), leading to n-3 lipophenol derivatives (Crauste, Rosell, Durand, & Vercauteren, 2016). Combining both therapeutic effects of n-3 PUFAs and phenolics in a lipophenolic molecule might be envisaged. Crauste et al. (2016) reviewed the synthesis and biological activities of n-3 lipophenols.

Enzymatic Lipophilization of Phenolics and Phenolic Extract

Zheng et al. (2010) prepared feruloylated lipids 1(3)-feruloyl-monobutyryl-glycerol (FMB) and 1(3)-feruloyl-dibutyryl-glycerol (FDB) by lipase-catalyzed transesterification. The reaction was performed in a closed tube containing Novozym 435, ethyl ferulate, tributyrin, and toluene. The reaction was incubated at 50 °C for 120 h. The enzyme was filtered then the filtrate was concentrated. Lucas et al. (2010) prepared hydroxytyrosol fatty acid esters by adding lipase (Novozym 435) to a mixture of hydroxytyrosol or tyrosol in t-butyl methyl ether. The mixture was stirred at 40 °C then the enzyme was separated.

Reddy, Ravinder, Prasad, and Kanjilal (2011) synthesized capsiate using lipase (Novozym 435) esterification of fatty acids with vanillyl alcohol. Fatty acid and vanillyl alcohol were solubilized in tertbutanol followed by the addition of lipase. The mixture was stirred at 55 °C then filtered to separate the lipase. The product was purified utilizing column chromatography with silica gel, and the structural confirmation was conducted. Laguerre et al. (2011) prepared chlorogenate esters using chemo-enzymatic esterification. Chlorogenic acid was dissolved in methanol. Amberlite IR 120 H was added and stirred at 55 °C. The mixture was filtered then methanol was removed. Chloroform was added, and the solution was dried and filtered. The resulted methyl chlorogenate was added to desired fatty alcohol, and the mixture was stirred at 55 °C till the dissolution of methyl chlorogenate. Lipase B was later added, and the suspension was heated (55 °C) under nitrogen. Lipophilized esters were purified using liquid-liquid extraction, and the alcohol were removed using column chromatography.

Wang and Shahidi (2014) synthesized monooleyl and dioleyl p-coumarates by the lipase-catalyzed reaction. The solution of p-coumaric acid was produced in

2-butanone, and a triolein stock solution was prepared in *n*-hexane. Volumes of *p*-coumaric acid and triolein stock solution were transferred to a flask followed by Novozyme addition. The reaction was performed at 50 °C then the solvent was removed under nitrogen. Durand et al. (2015) synthesized decyl rosmarinate mono-ester. Rosmarinic acid was solubilized in THF, then n-decanol and acidic sulfonic resin Amberlite1 IR-120H were added. The mixture was stirred at 57 °C. Aliquots were withdrawn from the mixture then mixed with methanol and filtered. Bayrasy et al. (2013) lipophilized RA by aliphatic chain lengths (i.e., butyl and octadecyl) to prepare rosmarinate alkyl esters.

Aladedunye et al. (2015) lipophilized *Sorbus aucuparia* extracts using lipase and octadecanol. The extract and octadecanol in 2-methyl-2-butanol was placed in a brown vessel. The molecular sieve was added, then the vessel was kept at 55 °C. Yarra et al. (2016) prepared (Z)-methyl-12-(methan sulfonyloxy) octadec-9-enoate, (Z)-methyl-12-(azido)octadec-9-enoate, and (Z)-methyl-12-(amino)octadec-9-enoate. Synthesis based on ricinoleic acid employing castor oil and phenolic acids. Pande and Akoh (2016) prepared lipase-catalyzed lipophilic tyrosyl esters with different degrees of unsaturation. Tyrosol was reacted with acyl donors in a 1:10 molar ratio (tyrosol: acyl donor). THF was added, then the substrates were mixed. Lipase and molecular sieves were added to eliminate moisture. Tubes were flushed with nitrogen, and the reaction was performed at 50 °C. Wang, Zhang, et al. (2016) prepared policosanyl phenolates by reacting vinyl phenolates with the Policosanols under Novozyme 435 in hexane/2-butanone at 60 °C for 4 days. A mixture of hexane, policosanols, vinyl 4-hydroxybenzoate, 2-butanone, and Novozyme was stirred under nitrogen. The mixture was filtered, the solution was evaporated, and the residue was purified using column chromatography.

Paximada, Echegoyen, Koutinas, Mandala, and Lagaron (2017) esterified EGCG with stearoyl chloride. The chloride was added to EGCG dissolved in ethyl acetate, in the presence of pyridine. The mixture was stirred, cooled, and filtrated. The filtrate was washed, and the organic layer was collected and evaporated to obtain a dry yellow powder. Liu and Yan (2019) lipophilized EGCG in the fast protocol. EGCG was added to acetone and heated at 40 °C. After dissolution, sodium acetate was added, and palmitoyl chloride was added to the solution. The reaction mixture was performed in a water bath under no nitrogen. Ethyl acetate was added to extract the mixture. The organic phase was washed, dried with anhydrous Na_2SO_4, and concentrated. Oh and Shahidi (2017) prepared resveratrol derivatives by esterification with acyl chlorides of fatty acids. The mono- and di-esters of crude products were utilized to assess their antioxidant effects in food and biological systems.

Characterization and Biological Properties of Lipo-phenolics

Table 3.2 presents the functional and biological activities of lipo-phenolics. Alemán et al. (2015) tested the antioxidant potential of lipophilized CA in fish oil-supplemented mayonnaise and milk. Alcohols-esterified CA were better

Table 3.2 Biological and functional activities of lipo-phenolics

Lipo-phenolic	Test type	Biological activity	References
Alcohols-esterified CA	Antioxidant activity in fish oil-enriched mayonnaise and milk	High antioxidant activity	Alemán et al. (2015)
Lipophilized dihydrocaffeic acid (octyl dihydrocaffeate and oleyl dihydrocaffeate)	Antioxidative activity in emulsions	Octyl dihydrocaffeate had a higher antioxidative effect than oleyl dihydrocaffeate in emulsions.	Sørensen, Nielsen, et al. (2012)
Lipo-phenolic modified from dihydrocaffeic acid and rutin.	Antioxidative activity in fish oil-enriched milk emulsions	Dihydrocaffeate esters and rutin laurate showed high antioxidant traits.	Sørensen, Petersen, et al. (2012)
CA and caffeates	Antioxidative activity in Citrem and Tween emulsions	CA was the most efficient antioxidant in the presence of tocopherol.	Sørensen et al. (2017)
Octyl esters of FA, sinapic acid, and CA	Rancimat method and DPPH· test. Antimicrobial tests	High antioxidant potential in rapeseed oil Octyl sinapate inhibited the growth of yeast, Gram-positive and Gram-negative bacteria	Szydłowska-Czerniak, Rabiej, Kyselka, et al. (2018)
Esterified triolein with cinnamic acid and FA	Antiradical activity	Lipophilized FA had higher radical scavenger impact than lipophilized cinnamic acid	Choo and Birch (2009)
Esterified flaxseed oil with cinnamic acid or FA (dicinnamoyl-monoacylglycerol, monocinnamoyl/feruloyldiacylglycerol, and monocinnamoyl-monoacylglycerol)	Antiradical activity	Lipophilized FA was more efficient radical quencher compared to lipophilized cinnamic acid	Choo et al. (2009)
Alkyl ferulates	Antioxidative activity in fish oil-enriched milk	Methyl ferulate was the most effective followed by FA and butyl ferulate	Sørensen et al. (2015)
Butyl ferulate	in vitro antioxidant and antimicrobial traits	Lipo-phenolic showed moderate antioxidant potential in DPPH· assay but exhibited good activity in linoleic acid oxidation test	Kaki, Kunduru, Kanjilal, and Prasad (2015)

(continued)

Table 3.2 (continued)

Lipo-phenolic	Test type	Biological activity	References
Lipophilic octyl sinapate	Antioxidant capacity (DPPH· and ABTS tests)	Octyl sinapate have antioxidant characters and could be added as antioxidants to edible oils	Szydłowska-Czerniak, Rabiej, and Krzemiński (2018)
Lipophilized *Sorbus aucuparia* extracts	Antioxidative activity in rapeseed oil at 65 °C and 180 °C	Rapeseed oil enriched with lipophilized extract showed high stability Lipophilized extract showed higher protection against degradation during frying	Aladedunye et al. (2015)
Lipophilized blueberry extracts	Antioxidative activity (stability of oil under high temperature)	Lipophilization of extracts improved oil stability	Luo et al. (2017)
Alkyl esters of protocatechuic acid	Anticancer tests	Increased toxicity Compounds with 8–14 carbons being the most toxic and displaying IC_{50} in the nanomolar range	Pereira et al. (2017)
Hydroxytyrosyl esters	Antioxidant potential in phosphatidylcholine liposomes	Lipophilic chain length have effects on the antioxidant effects	Balducci, Incerpi, Stano, and Tofani (2018)
1,2-diacylglycerol rosmarinate (RDAG) with different alkyl chain lengths	ROS overexpressing fibroblasts	RDAG12 displayed the best antioxidant potential	Durand et al. (2015)
RA alkyl esters	Membrane interactions in liposomes	Differences were noted between lipo-phenolics with respect to their interaction and location within the bilayer	Durand, Jacob et al. (2017)
Rosmarinate alkyl esters	Antioxidant potential (level of ROS)	Increasing the chain length led to an improvement of the antioxidant activity until a threshold is reached (10 carbon atoms)	Bayrasy et al. (2013)

Coupling (Z)-methyl-12-aminooctadec-9-enoate with phenolic acids	Antioxidant potential (DPPH·, lipid peroxidation inhibitory activity, superoxide free radical scavenging activity) Anticancer activity	High antioxidant activity Compounds from cinnamic acid-based phenolic acids had high anticancer activity	Yarra et al. (2016)
Tyrosyl esters with carbon alkyl chain and different degrees of unsaturation	Antioxidant traits in an o/w emulsion and oil-structured lipid	Tyrosyl oleate was the most efficient antioxidant in the emulsion followed by tyrosyl stearate and tyrosyl linoleate	Pande and Akoh (2016)
FMB and FDB	Antioxidant potential	Antioxidant properties decreased in the following order: BHT > FMB > FDB > FA	Zheng et al. (2010)
Lipophilization of phenolates using policosanols	Antioxidant potential (ABTS and linoleic acid peroxidation ferric thiocyanate assays)	Policosanyl syringate, policosanyl 4-hydroxybenzoate, and policosanyl 4-hydroxyphenylacetate showed high inhibition effects on lipid oxidation	Wang, Hwang, Lim (2016)
Lipophilized catechin (L-EGCG)	Antioxidant potential, emulsion stability, droplet size, bulk, and interfacial viscosity	Ultrasound homogenization combined with L-EGCG was the most adequate combination	Paximada et al. (2017)
Lipophilized epicatechin with FFA	Antioxidant potential in oils	Lipophilized epicatechin enhanced the oxidative stability of oils	Chen et al. (2017)
Lipophilized EGCG	Antiradical potential	The length of fatty acid chain positively correlated with antiradical capacity	Wang et al. (2016)
Lipophilized EGCG	Antiradical potential	High radical scavenging capacities	Liu and Yan (2019)
Hydroxytyrosol fatty acid esters	Antioxidant potential in fish o/w emulsions	Hydroxytyrosol octanoate had the highest antioxidant potential	Lucas et al. (2010)
Salidroside liposomes	Physicochemical stability	Salidroside liposomes exhibited the slower increase in particle size than liposomes without salidroside	Fan, Xu, Xia, and Zhang (2007)

(continued)

Table 3.2 (continued)

Lipo-phenolic	Test type	Biological activity	References
Lipophilized capsiate	Antioxidant capacity (DPPH· and Rancimat assays)	DPPH· and Rancimat assays did not show specific trend of antioxidant activity	Reddy et al. (2011)
Chlorogenic acid esters	ROS-overexpressing fibroblast cell line	Dodecyl and hexadecyl esters were more active than methyl, butyl, and octyl esters	Laguerre et al. (2011)
Lipophilized *p*-coumaric acid with triolein	Antioxidant capacity (peroxyl radical scavenging, DPPH· and reducing power, β-carotene/linoleate bleaching, LDL cholesterol oxidation, peroxyl radical-induced DNA cleavage assays)	Lipo-phenolics exhibited different antioxidant traits	Wang and Shahidi (2014)
Lipophilic quercetin esters with oleic, linoleic and linolenic acids	Antioxidant activity	Different antioxidant activities	Mainini et al. (2013)
Lipophilized quercetin, carvacrol and vanillin	Antioxidant potential	Lipophilized products had high lipophilicity and higher antioxidant potential	Ortega-Valencia et al. (2019)
Lipophilized resveratrol	Antioxidant activity in o/w emulsion and ground meat	Resveratrol esters showed high antioxidant potential in ground meat	Oh and Shahidi (2018)
Dopamine esters	Antioxidant activity in oils Hemolytic activity against human erythrocyte	Long chain fatty acid esters displayed a high protective effect of oil Middle chain length derivatives showed high hemolytic activity against human erythrocyte	Sellami et al. (2013)
Chlorogenic acid esters	Antiradical activity	Improved antiradical activity	López-Giraldo et al. (2009)

antioxidants than the CA. Fish oil- supplemented mayonnaise with caffeates of medium alkyl chain length resulted in higher stability than caffeates with longer or shorter alkyl chains. In fish oil- supplemented milk emulsions, the most active caffeates were shorter alkyl chains rather than medium and long chains. The findings demonstrated that there might be an optimum alkyl chain length for each lipo-phenolic in each emulsion system.

Sørensen, Nielsen, et al. (2012) tested the antioxidant potentail of lipophilized dihydrocaffeic acid (octyl dihydro caffeate and oleyl dihydro caffeate). Figure 3.1 presents the structure of CA, and dihydrocaffeic acid. Despite low antioxidant traits of lipophilized dihydrocaffeic acid in different antioxidant tests, lipophilized dihydrocaffeic acid was more active than CA and dihydrocaffeic acid. Octyl dihydro caffeate had a better antioxidant impact than oleyl dihydro caffeate in the emulsion systems. The results supported the polar paradox hypothesis since lipophilized products had higher stability.

On the other hand, the decreased antioxidant effect with increasing alkyl chain length esterified to dihydrocaffeic acid supported a cutoff effect theory. The result is suggested to explain the low antioxidative action of oleyl dihydro caffeate compared with octyl dihydro caffeate.

Sørensen, Petersen, et al. (2012) tested the antioxidant impact of lipo-phenolics in fish oil- supplemented milk emulsions. Two lipo-phenolic modified from dihydrocaffeic acid and rutin were tested. Rutin laurate and dihydrocaffeate esters exhibited higher antioxidant traits in the emulsion system compared with the pure phenolics. Rutin palmitate exhibited higher antioxidant potential than rutin. The rutin results supported a cut-off effect on the alkyl chain length concerning optimal antioxidant effect in the emulsion systems. Therefore, the optimum alkyl chain length could be below 16 C atoms and might be less for rutin esters. Sørensen et al. (2017) evaluated the effect of emulsifiers (Citrem and Tween 80) and the presence

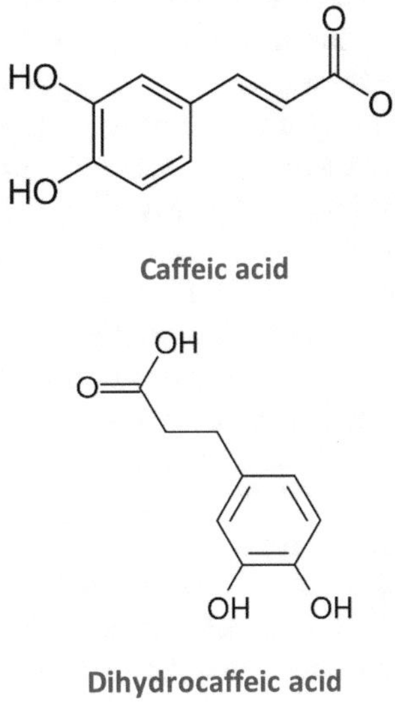

Fig. 3.1 Structure of caffeic acid, and dihydrocaffeic acid

Caffeic acid

Dihydrocaffeic acid

of tocols on the efficacies CA and caffeates (C1-C20) in the emulsion systems. Oxidation was tested during storage, and partitioning of CA and caffeates was determined in the water phase. Partitioning of CA and caffeates was affected by the presence of tocols and emulsifier type. CA was the most effective antioxidant in Citrem, and Tween stabilized the emulsions in the presence of tocols. On the other hand, for Tween stabilized emulsion, CA acted as a pro-oxidant, and the caffeates exhibited strong antioxidative effects in the absence of tocols. The differences noted in antioxidant potential with different emulsifiers and tocopherols are due to the antioxidant-antioxidant and emulsifier–antioxidant interaction in the emulsion.

Szydłowska-Czerniak, Rabiej, Kyselka, et al. (2018) added octyl esters of FA, sinapic acid, and CA, to rapeseed oil as antioxidants. The impact of octyl esters on the stability of rapeseed oil was studied using the Rancimat test, wherein a positive linear correlation between the induction times (IP) and lipo-phenolic level were noted. The antiradical effect of oils with lipo-phenolic was tested against DPPH· radicals. Enriched rapeseed oils showed higher antioxidant potential and an increase in the oil stability against oxidation. Besides, octyl sinapate inhibited the growth of Gram-positive and Gram-negative bacteria as well as the yeast.

Choo and Birch (2009) conducted lipase-catalyzed trans-esterification of triolein with FA and cinnamic acid. FA showed higher antiradical action than cinnamic acid. The esterification of cinnamic acid and FA with triolein casued an increase and a decrease in the antiradical action, respectively. Lipophilized FA had higher radical quencher impact than lipophilized cinnamic acid. The impact of esterification of cinnamic acid was tested using ethyl cinnamate, which improves the antiradical action. The antiradical action of unseparated mixture of lipophilized products was as good as monocinnamoyl dioleoyl glycerol. Choo et al. (2009) also conducted lipase-catalyzed trans-esterification of flaxseed oil with cinnamic acid or FA. Lipase-catalyzed transesterification of flaxseed oil with cinnamic acid or FA formed lipo-phenolic products (monocinnamoyl/feruloyldiacylglycerol, and monocinnam-oyl-monoacylglycerol). The esterification of cinnamic acid or FA with oil caused a high increase and decrease in the antiradical action compared with pure phenolic acids, respectively. Lipophilized FA was a more active radical quencher compared to lipophilized cinnamic acid and provided improved antioxidative potentail in the flaxseed oil. Sørensen et al. (2015) evaluated the antioxidant potential of unbranched alkyl ferulates in fish oil-enriched milk. A cut-off effect in the antioxidant potential concerning the alkyl chain length was noted. Methyl ferulate was the most effective followed by FA and butyl ferulate, whereas octyl ferulate was pro-oxidative and the pro-oxidative impact increased with the increase in the alkyl chain length.

Further elongation of the alkyl chain length to C16 and C20 caused a weak pro-oxidative impacts to weak antioxidant impact. Kaki et al. (2015) modified FA to form a lipo-phenolic containing butyl chain. FA esterified with butanol to form butyl ferulate taht was dihydroxylated followed by esterification with butyric anhy-dride to form lipo-phenolic containing butyric acid. NMR, IR, and MS methods approved the structure of the structured lipo-phenolic. The structured lipo-phenolic was *in vitro* tested for antimicrobial and antioxidant traits. Lipo-phenolic exhibited moderate antioxidative effects in DPPH· test, but in the linoleic acid oxidation assay, it showed good effect. The lipo-phenolic might find uses in lipophilic systems and also as a novel prodrug of butyric acid. In addition, it exhibited an antibacterial potential against some bacterial strains.

Szydłowska-Czerniak, Rabiej, and Krzemiński (2018)prepared lipophilic octyl sinapate, and chromatographic and spectroscopic analyses confirmed the synthesized lipo-phenolics. The antioxidant effect of lipo-phenolics was measured by DPPH· and ABTS tests. The antioxidative effect of rapeseed-linseed oil contained the esters was about 30 times greater in comparison with the control oil sample. The octyl caffeate, octyl sinapate, and octyl ferulate showed antioxidant and lipophilic characters; thus, they might be used as antioxidants in edible fats and oils.

Aladedunye et al. (2015) lipophilized *Sorbus aucuparia* extracts using lipase and octadecanol as the alkyl donor. The antioxidative potential of lipophilized extracts was tested in rapeseed oil at 65 °C and under frying conditions at 180 °C. The oil enriched with lipophilized extracts showed better oxidative stability with a decrease in peroxide value (PV). During frying, lipophilized extract exhibited better protection against degradation. At the end of frying experiments, French fries processed in the presence of lipophilized extracts had more phenolics, which indicate better solubilization of phenolics under frying conditions. Luo et al. (2017) added blueberry anthocyanin extract to camellia oil and lipophilized with FFA in the oil through the enzymatic reaction. Enzymatic lipophilization of extracts with FFA improved the stability of the oil under high temperatures, while the FFA level in camellia seed oil was decreased.

Pereira et al. (2017) evaluated lipo-phenolics containing alkyl esters derivatives of protocatechuic acid against cancer and non-cancer cell lines. The modifications by the inclusion of carbon side chains caused a high toxicity, wherein thecompounds having 8–14 carbons being the most toxic, and showing IC_{50} in the nanomolar range. Balducci et al. (2018) measured the functional traits and the antioxidative potential of hydroxytyrosyl esters in phosphatidylcholine (PC) liposomes. The distribution of hydroxytyrosyl long-chain esters depend on their lipophilic chain length, and this turns to have impacts on their antioxidative properties.

Durand et al. (2015) synthesized lipo-phenolics derived from RA (Fig. 3.2a) including 1,2-diacylglycerol rosmarinate (RDAG) with different alkyl chain lengths. The antioxidative capabilities of lipo-phenolics were tested on ROS overexpressing fibroblasts. RDAG12 showed the highest antioxidant potential. RDAG12

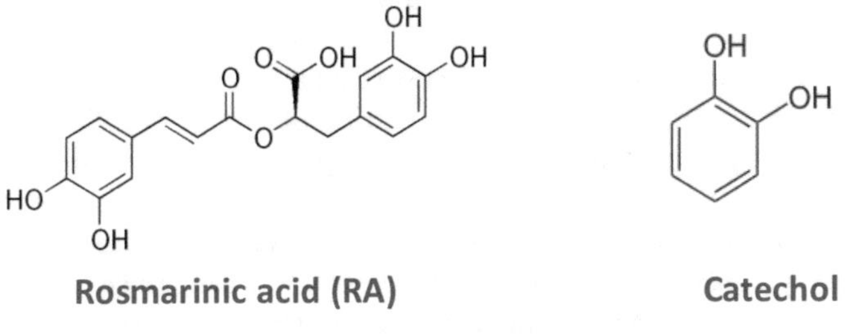

Rosmarinic acid (RA)   **Catechol**

Fig. 3.2 Structure of rosmarinic acid (RA) and catechol

antioxidant potential was compared with vitamins E and C, RA, and decyl rosmarinate (R10). The order of antioxidant activity was R10 > RDAG12 > RDAG5 > RA > vitamin C > vitamin E. Catechol structure (Fig. 3.2b) is active in terms of antioxidative action and linking a hydrophobic domain is a novel was to form active antioxidants. Durand, Jacob, et al. (2017) and Durand, Lecomte, et al. (2017) investigated the membrane interaction of RA, and its alkyl esters in model liposomes. Differences were noted between lipo-phenolics concerning their interaction and location within the bilayer. Both penetration depth and membrane affinity follow a nonlinear behavior. The X-ray scattering analyses were correlated with lipo-phenolics' properties noted in the biological system and lipid dispersions. Bayrasy et al. (2013) lipophilized RA by different aliphatic chain lengths to give rosmarinate alkyl esters that were tested for their capability to reduce ROS levels. Increasing the chain length casue an improvement of the antioxidative potential activity till a threshold is reached for medium-chain.

Yarra et al. (2016) synthesized lipo-phenolics by coupling (Z)-methyl-12-aminooctadec-9-enoate with phenolic acids. Lipo-phenolics were characterized by FT-IR, ESI-MS, 1H NMR, 13C NMR, and HRMS. *in vitro* antioxidative properties of lipo-phenolics were also tested by DPPH·, lipid peroxidation inhibitory effect and superoxide free radical scavenging effect. The lipo-phenolics were compared in terms of antioxidative potential with the α-tocopherol and BHT. The antioxidative results of the lipo-phenolics showed that DPPH· antiradical effect, superoxide, and lipid peroxidation inhibitory potential of CA and gallic acid substituted lipophenolics showed high effects. Lipo-phenolics were screened for anticancer traits against cancer cell lines, wherein the compounds from cinnamic acid-based acids showed high anticancer activity.

Pande and Akoh (2016) prepared lipophilic tyrosyl esters with a carbon alkyl chain and different unsaturation degrees. FFA and fatty acid ethyl esters used as acyl donors and lipase was the biocatalyst. The lipo-phenolics were recovered and characterized. Peroxide and *p*-anisidine values were determined to evaluate their antioxidant traits in an o/w emulsion and oil-structured lipid. The reaction time increased, as the unsaturation of fatty acid increased. In oil-structured lipid, tyrosyl esters showed low antioxidant potential, wherein an alkyl chain improved the antioxidative impact of tyrosol. Tyrosyl oleate was the main active antioxidant in the emulsion, followed by tyrosyl stearate then tyrosyl linoleate.

Zheng et al. (2010) studied the structures of FDB and FMB using ESI-MS and NMR. The antioxidant properties decreased following the order: BHT > FMB > FDB > FA. FDB and FMB showed high antioxidant effects in lipophilic system, making them novel antioxidant compounds. Wang, Hwang, et al. (2016) and Wang, Zhang, et al., (2016) carried out enzymatic lipophilization of phenolates using policosanols. The antioxidant potential of the resulting lipo-phenolics was compared using ABTS and linoleic acid per-oxidation ferric thiocyanate assays. Policosanyl phenolates exhibited a lower ABTS antiradical effect. Policosanyl syringate, policosanyl 4-hydroxybenzoate, and policosanyl 4-hydroxyphenylacetate showed high inhibition impacts on lipid oxidation.

Paximada et al. (2017) studied the potential of emulsion electrospraying that contained protein and cellulose for EGCG encapsulation. Hydrophilic catechin or lipophilized catechin (L-EGCG) were encapsulated to study antioxidant stability. Droplet size, emulsion stability, bulk, and interfacial viscosity were tested. Ultrasound homogenization, combined with L-EGCG, proved to be a good combination. Chen et al. (2017) reduced the FFA in camellia seed oil by lipophilization of epicatechin with FFA. The oil acid value was reduced after lipophilization. Epicatechin palmitate and epicatechin oleate were formulated in the lipophilized oil. The PV, *p*-anisidine value, and total oxidation values during thermal treatment of lipophilized oil were lower than that of the oil, confirming that lipophilized epicatechin might improve the oil oxidative stability. Wang, Zhang, et al. (2016) prepared lipophilized EGCG derivatives by EGCG esterification with aliphatic fatty acids. The length of the fatty acid chain positively correlated with antiradical activity, but bulky long-chain substitutes prevented methylglyoxal trapping. Lipophilization, except for docosahexaenoic acid, exhibited no interference with EGCG's *in vitro* effect of advanced glycation end-products formation. Liu and Yan (2019) developed a rapid technique to synthesize lipophilic EGCG derivatives. HPLC-MS confirmed monoesters of EGCG derivatives, and the predominant product was 4′-O-palmitoyl EGCG. The lipophilized EGCG exhibited high antiradical effects. Compared to EGCG, the solubility of lipophilized EGCG was enhanced 470 times in lard.

Lucas et al. (2010) noted a nonlinear behavior in the antioxidative potential of hydroxytyrosol fatty acid esters in fish o/w emulsion, wherein hydroxytyrosol octanoate showed the highest antioxidant potential. Hydroxytyrosol and tyrosol fatty acid esters were suitable surfactants when the right hydrophilic-lipophilic balance (HLB) was attained and as effective as emulsifiers utilized in the industry. A nonlinear dependency of surfactant activity was noted with the increase in chain length of the lipophilic antioxidant. Fan et al. (2007) prepared salidroside liposomes. A straight-line leakage tendency of salidroside was noted. Salidroside liposomes had higher physico-chemical stability. When the temperature increased, instability in the systems was exacerbated. Salidroside liposomes showed a slower increase in particle size than liposomes without salidroside.

Reddy et al. (2011) synthesized capsiate by esterification of vanillyl alcohol with fatty acids using lipase. DPPH· and Rancimat tests did not exhibit a specific trend of antioxidant potential with the increase in lipophilicity and the type of fatty acid grafted to the phenolic moieties. Laguerre et al. (2011) evaluated the antioxidative capacity of chlorogenic acid, and its esters toward mitochondrial ROS generated in a ROS-overexpressing fibroblast cell line. Dodecyl and hexadecyl esters were more active than methyl, butyl, and octyl esters. Dodecyl chlorogenate showed high antioxidative potential, for incubation time and concentration below the cytotoxicity threshold. Wang and Shahidi (2014) conducted lipase-catalyzed acidolysis of *p*-coumaric acid with triolein. The identification of structured lipo-phenolics was performed using HPLC-MS. Lipo-phenolics exhibited different antioxidant traits in peroxyl radical quenching, DPPH· and reducing power, human LDL-cholesterol oxidation, β-carotene/linoleate bleaching, as well as peroxyl radical-induced DNA cleavage assays.

Mainini et al. (2013) synthesized lipophilic quercetin esters with linoleic, oleic and linolenic acids. Tetraesters and triesters were obtained by modulating the reaction condition. The antioxidant activity of synthesized lipo-phenolics was determined. Ortega-Valencia et al. (2019) lipophilized quercetin, carvacrol, and vanillin by a lipophilic reaction with CLA. 1H NMR analyzed the structure of the precursor and lipophilized compounds. Carvacrol-CLA and quercetin-CLA showed the highest lipophilization yields. The lipophilized systems had higher lipophilicity and higher antioxidant potential than the phenolic precursors, wherein vanillin-CLA and carvacrol-CLA showed the greatest antioxidant potential.

Oh and Shahidi (2018) lipophilized resveratrol (R) using acyl chlorides of different chain lengths. Resveratrol displayed the greatest antioxidative potential in o/w emulsion, but its lipophilized esters showed higher antioxidant activity in the oil system. Resveratrol esters R-C20:5n-3 and R-C22:6n-3 displayed greatest antioxidative antioxidant potential in ground meat. Resveratrol derivatives (R-C3:0–R-C14:0) had higher hydrogen peroxide scavenging activity than R.

Sellami et al. (2013) prepared dopamine esters with high lipophilicity using lipase. The best conversion yield was achived when caprylic acid was utilized as an acyl donor. Long-chain fatty acid esters showed a higher protective impact of oil against oxidation than dopamine or BHT. Medium-chain length derivatives of oleic acid derivative and dopamine showed the best hemolytic effect against human erythrocyte. López-Giraldo et al. (2009) studied the antiradical traits of chlorogenic acid (5-CQA) and its esters. Hydrophobation improved the antiradical potential of 5-CQA esters. It seems that the pathways of DPPH· stabilization were different between 5-CQA and its esters.

Conclusion and Further Research

Lipo-phenolics present exciting potential for extending their applications from lipophilic antioxidant compounds towards functional molecules. Combining the effects of their polar head with their capability to interfere with membranes is conveyed by their lipophilic tail. Lipo-phenolics could be exploited in several lipophilic bioactive compounds and health-promoting formulations. The high lipophilicity and antioxidant potential of lipo-phenolics allow them as novel stabilizers of lipid-containing food against oxidation. Lipo-phenolics could be added to the deep-fried products with the potential to enhance storage stability and nutritional quality. Using of active lipo-phenolics might reduce the amount of antioxidants required to protect products against lipid oxidation which could reduce the cost.

Lipophilization of phenolics improves their antioxidant and other biological traits. The optimal alkyl chain length for lipo-phenolics depends upon the studied matrix. The enzymatic process to lipophilize phenolics could increase the antioxidant activity and obtain functional products with high stability. Different factors may affect the esterification, including the enzyme selectivity and stability, the solubility of substrates and products, solvent polarity, water content, and temperature.

The raw materials used for the formulation of lipo-phenolics are natural compounds; however, the fate of the lipo-phenolics *in vivo* must be tested. Research under different conditions is important to study *in vitro* release and *in vivo* availability of lipo-phenolics. Further studies on the toxicological and functional traits of lipo-phenolics should be performed, followed by an economic yield under industrial scales.

References

Aladedunye, F., Niehaus, K., Bednarz, H., Thiyam-Hollander, U., Fehling, E., & Matthäus, B. (2015). Enzymatic lipophilization of phenolic extract from rowanberry (*Sorbus aucuparia*) and evaluation of antioxidative activity in edible oil. *LWT – Food Science and Technology, 60*(1), 56–62. https://doi.org/10.1016/j.lwt.2014.08.008

Alemán, M., Bou, R., Guardiola, F., Durand, E., Villeneuve, P., Jacobsen, C., et al. (2015). Antioxidative effect of lipophilized caffeic acid in fish oil enriched mayonnaise and milk. *Food Chemistry, 167*, 236–244. https://doi.org/10.1016/j.foodchem.2014.06.083

Anselmi, C., Centini, M., Andreassi, M., Buonocore, A., La Rosa, C., Maffei Facino, R., et al. (2004). Conformational analysis: A tool for the elucidation of the antioxidant properties of ferulic acid derivatives in membrane models. *Journal of Pharmaceutical and Biomedical Analysis, 35*, 1241–1249.

Aruoma, O. I., Murcia, A., Butler, J., & Halliwell, B. (1993). Evaluation of the antioxidant and pro-oxidant actions of gallic acid and its derivatives. *Journal of Agricultural and Food Chemistry, 41*, 1880–1885.

Balakrishna, M., Kaki, S. S., Karuna, M. S. L., Sarada, S., Kumar, F. G., & Prasad, R. B. N. (2017). Synthesis and in vitro antioxidant and antimicrobial studies of novel structured phosphatidylcholines with phenolic acids. *Food Chemistry, 221*, 664–672. https://doi.org/10.1016/j.foodchem.2016.11.121

Balducci, V., Incerpi, S., Stano, P., & Tofani, D. (2018). Antioxidant activity of hydroxytyrosyl esters studied in liposome models. *Biochimica et Biophysica Acta-Biomembranes, 1860*(2), 600–610. https://doi.org/10.1016/j.bbamem.2017.11.012

Bayrasy, C., Chabi, B., Laguerre, M., Lecomte, J., Jublanc, É., Villeneuve, P., et al. (2013). Boosting antioxidants by lipophilization: A strategy to increase cell uptake and target mitochondria. *Pharmaceutical Research, 30*(8), 1979–1989. https://doi.org/10.1007/s11095-013-1041-4

Bernini, R., Crisante, F., Barontini, M., Tofani, D., Balducci, V., & Gambacorta, A. (2012). Synthesis and structure/antioxidant activity relationship of novel catecholic antioxidant structural analogues to hydroxytyrosol and its lipophilic esters. *Journal of Agricultural and Food Chemistry, 60*, 7408–7416.

Bernini, R., Mincione, E., Barontini, M., & Crisante, F. (2008). Convenient synthesis of hydroxytyrosol and its lipophilic derivatives from tyrosol or homovanillyl alcohol. *Journal of Agricultural and Food Chemistry, 56*, 8897–8904.

Buisman, G. J. H., van Helteren, C. T. W., Kramer, G. F. H., Veldsink, J. W., Derksen, J. T. P., & Cuperus, F. P. (1998). Enzymatic esterifications of functionalized phenols for the synthesis of lipophilic antioxidants. *Biotechnology Letters, 20*, 131–136.

Chalas, J., Claise, C., Edeas, M., Messaoudi, C., Vergnes, L., Abella, A., et al. (2001). Effect of ethyl esterification of phenolic acids on low-density lipoprotein oxidation. *Biomedicine & Pharmacotherapy, 55*, 54–60.

Cheetham, P. S. J., & Banister, N. E. (2000, May 23). Production and uses of caffeic acid and derivatives thereof. U.S. Patent 6, 066,311, 19 pp.

Chen, S. S., Luo, S. Z., Zheng, Z., Zhao, Y. Y., Pang, M., & Jiang, S. T. (2017). Enzymatic lipophilization of epicatechin with free fatty acids and its effect on antioxidative capacity in crude camellia seed oil. *Journal of the Science of Food and Agriculture, 97*(3), 868–874. https://doi.org/10.1002/jsfa.7808

Choo, W. S., & Birch, E. J. (2009). Radical scavenging activity of lipophilized products from lipase-catalyzed transesterification of triolein with cinnamic and ferulic acids. *Lipids, 44*(2), 145–152. https://doi.org/10.1007/s11745-008-3242-x

Choo, W. S., Birch, E. J., & Stewart, I. (2009). Radical scavenging activity of lipophilized products from transesterification of flaxseed oil with cinnamic acid or ferulic acid. *Lipids, 44*(9), 807–815. https://doi.org/10.1007/s11745-009-3334-2

Compton, D. L., Laszlo, J. A., & Berhow, M. A. (2000). Lipase-catalyzed synthesis of ferulate esters. *Journal of the American Oil Chemists' Society, 77*, 513–519.

Cottrell, T., & van Peij, J. (2004). Sorbitan esters and polysorbates. In R. J. Whitehurst (Ed.), *Emulsifiers in food technology* (pp. 162–185). Hoboken, NJ: Wiley-Blackwell.

Crauste, C., Rosell, M., Durand, T., & Vercauteren, J. (2016). Omega-3 polyunsaturated lipophenols, how and why? *Biochimie, 120*, 62–74. https://doi.org/10.1016/j.biochi.2015.07.018

Cuvelier, M.-E., Richard, H., & Berset, C. (1992). Comparison of the antioxidative activity of some acid-phenols: Structure-activity relationship. *Bioscience, Biotechnology, and Biochemistry, 26*, 324–325.

Decker, E. A. (1998). Strategies for manipulating the prooxidative/antioxidative balance of foods to maximize oxidative stability. *Trends in Food Science and Technology, 9*, 241–248.

Durand, E., Bayrasy, C., Laguerre, M., Barouh, N., Lecomte, J., Durand, T., et al. (2015). Regioselective synthesis of diacylglycerol rosmarinates and evaluation of their antioxidant activity in fibroblasts. *European Journal of Lipid Science and Technology, 117*(8), 1159–1170. https://doi.org/10.1002/ejlt.201400607

Durand, E., Jacob, R. F., Sherratt, S., Lecomte, J., Baréa, B., Villeneuve, P., et al. (2017). The nonlinear effect of alkyl chain length in the membrane interactions of phenolipids: Evidence by X-ray diffraction analysis. *European Journal of Lipid Science and Technology, 119*(8), 1–7. https://doi.org/10.1002/ejlt.201600397

Durand, E., Lecomte, J., & Villeneuve, P. (2017). The Biological and Antimicrobial activities of Phenolipids. *Lipid Technology, 29*(7–8), 67–70. https://doi.org/10.1002/lite.201700019

Fan, M., Xu, S., Xia, S., & Zhang, X. (2007). Effect of different preparation methods on physicochemical properties of salidroside liposomes. *Journal of Agricultural and Food Chemistry, 55*(8), 3089–3095. https://doi.org/10.1021/jf062935q

Fang, X., Shima, M., Kadota, M., Tsuno, T., & Adachi, S. (2006). Suppressive effect of alkyl ferulate on the oxidation of microencapsulated linoleic acid. *European Journal of Lipid Science and Technology, 108*, 97–102.

Figueroa-Espinoza, M. C., Laguerre, M., Villeneuve, P., & Lecomte, J. (2013). From phenolics to phenolipids: Optimizing antioxidants in lipid dispersions. *Lipid Technology, 25*, 131–134.

Figueroa-Espinoza, M. C., & Villeneuve, P. (2005). Phenolic enzymatic lipophilization. *Journal of Agricultural and Food Chemistry, 53*(8), 2779–2787.

Fiuza, S. M., Gomes, C., Teixeira, L. J., Girão da Cruz, M. T., Cordeiro, M. N. D. S., Milhazes, N., et al. (2004). Phenolic acid derivatives with potential anticancer properties – A structure-activity relationship study. Part 1: Methyl, propyl and octyl esters of caffeic and gallic acids. *Bioorganic & Medicinal Chemistry, 12*, 3581–3589.

Frankel, E. N., & Meyer, A. S. (2000). The problems of using onedimensional methods to evaluate multifunctional food and biological antioxidants. *Journal of the Science of Food and Agriculture, 80*, 1925–1941.

Fujita, K., & Kubo, I. (2002). Antifungal activity of octyl gallate. *International Journal of Food Microbiology, 79*, 193–201.

Giuliani, S., Piana, C., Setti, L., Hochkoeppler, A., Pifferi, P. G., Williamson, G., et al. (2001). Synthesis of pentylferulate by a feruloyl esterase from *Aspergillus niger* using water-in-oil microemulsions. *Biotechnology Letters, 23*, 325–330.

Guyot, B., Bosquette, B., Pina, M., & Graille, J. (1997). Esterification of phenolic acids from green coffee with an immobilized lipase from *Candida antarctica* in solvent-free medium. *Biotechnology Letters, 19*, 529–532.

Guyot, B., Gueule, D., Pina, M., Graille, J., Farines, V., & Farines, M. (2000). Enzymatic synthesis of fatty esters in 5-caffeoyl quinic acid. *European Journal of Lipid Science and Technology*, 102, 93–95. https://doi.org/10.1002/(SICI)1438-9312(200002)102:2 <93::AID-EJLT93>3.0.CO;2-B

Ha, T. J., Nihei, K.-I., & Kubo, I. (2004). Lipoxygenase inhibitory activity of octyl gallate. *Journal of Agricultural and Food Chemistry, 52*, 3177–3181.

Halliwell, B., Aeschbach, R., Loliger, J., & Aruoma, O. I. (1995). The characterization of antioxidants. *Food and Chemical Toxicology, 33*, 601–617.

Hills, G. (2003). Industrial use of lipases to produce fatty acid esters. *European Journal of Lipid Science and Technology, 105*, 601–607.

Jayaprakasam, B., Vanisree, M., Zhang, Y., Dewitt, D. L., & Nair, M. G. (2006). Impact of alkyl esters of caffeic and ferulic acids on tumor cell proliferation, cyclooxygenase enzyme, and lipid peroxidation. *Journal of Agricultural and Food Chemistry, 54*, 5375–5381.

Kaki, S. S., Kunduru, K. R., Kanjilal, S., & Prasad, R. B. N. (2015). Synthesis and characterization of a novel phenolic lipid for use as potential lipophilic antioxidant and as a prodrug of butyric acid. *Journal of Oleo Science, 64*(8), 845–852. https://doi.org/10.5650/jos.ess15035

Khmelnitsky, Y. L., Hilhorst, R., & Veeger, C. (1988). Detergentless microemulsions as media for enzymatic reactions. Cholesterol oxidation catalyzed by cholesterol oxidase. *European Journal of Biochemistry, 176*, 265–271.

Kikuzaki, H., Hisamoto, M., Hirose, K., Akiyama, K., & Taniguchi, H. (2002). Antioxidant properties of ferulic acid and its related compounds. *Journal of Agricultural and Food Chemistry, 50*, 2161–2168.

Kubo, I., Masuoka, N., Xiao, P., & Haraguchi, H. (2002). Antioxidant activity of dodecyl gallate. *Journal of Agricultural and Food Chemistry, 50*, 3533–3539.

Kubo, I., Xiao, P., & Fujita, K. (2001). Antifungal activity of octyl gallate: Structural criteria and mode of action. *Bioorganic & Medicinal Chemistry Letters, 11*, 347–350.

Laguerre, M., Bayrasy, C., Lecomte, J., Chabi, B., Decker, E. A., Wrutniak-Cabello, C., et al. (2013). How to boost antioxidants by lipophilization? *Biochimie, 95*, 20–26.

Laguerre, M., Bayrasy, C., Panya, A., Weiss, J., McClements, D. J., Lecomte, J., et al. (2013). What makes good antioxidants in lipid-based systems? The next theories beyond the polar paradox. *Critical Reviews in Food Science and Nutrition*. https://doi.org/10.1080/1040839 8.2011.650335

Laguerre, M., López Giraldo, L. J., Lecomte, J., Figueroa-Espinoza, M.-C., Baréa, B., et al. (2009). Chain length affects antioxidant properties of chlorogenate esters in emulsion: The cutoff theory behind the polar paradox. *Journal of Agricultural and Food Chemistry, 57*(23), 11335–11342.

Laguerre, M., López Giraldo, L. J., Lecomte, J., Figueroa-Espinoza, M.-C., Baréa, B., et al. (2010). Relationship between hydrophobicity and antioxidant ability of "phenolipids" in emulsion: A parabolic effect of the chain length of rosmarinate esters. *Journal of Agricultural and Food Chemistry, 58*(5), 2869–2876.

Laguerre, M., López-Giraldo, L. J., Lecomte, J., Baréa, B., Cambon, E., Tchobo, P. F., et al. (2008). Conjugated autoxidizable triene (CAT) assay: A novel spectrophotometric method for determination of antioxidant capacity using triacylglycerol as ultraviolet probe. *Analytical Biochemistry, 380*(2), 282–290.

Laguerre, M., Wrutniak-Cabello, C., Chabi, B., López Giraldo, L. J., Lecomte, J., Villeneuve, P., et al. (2011). Does hydrophobicity always enhance antioxidant drugs? A cut-off effect of the chain length of functionalized chlorogenate esters on ROS-overexpressing fibroblasts. *Journal of Pharmacy and Pharmacology, 63*(4), 531–540. https://doi. org/10.1111/j.2042-7158.2010.01216.x

Lecomte, J., Giraldo, L. J. L., Laguerre, M., Barea, B., & Villeneuve, P. (2010). Synthesis, characterization, free radical scavenging properties of rosmarinic acid fatty esters. *Journal of the American Oil Chemists' Society, 87*, 615–620.

Liu, B., & Yan, W. (2019). Lipophilization of EGCG and effects on antioxidant activities. *Food Chemistry, 272*, 663–669. https://doi.org/10.1016/j.foodchem.2018.08.086

López Giraldo, L. J., Laguerre, M., Lecomte, J., FigueroaEspinoza, M.-C., et al. (2007). Lipase-catalyzed synthesis of chlorogenate fatty esters in solvent-free medium. *Enzyme and Microbial Technology, 41*, 721–726.

Lopez-Giraldo, J. L., Laguerre, M., Lecomte, J., Figueroa-Espinoza, M.-C., Pina, M., & Villeneuve, P. (2007). Lipophilisation de composés phénoliques par voie enzymatique et propriétés anti-oxydantes des molécules lipophilisées. *Oléagineux, Corps Gras, Lipides, 14*(1), 51–59. https:// doi.org/10.1051/ocl.2007.0100

López-Giraldo, L. J., Laguerre, M., Lecomte, J., Figueroa-Espinoza, M. C., Baréa, B., Weiss, J., et al. (2009). Kinetic and stoichiometry of the reaction of chlorogenic acid and its alkyl esters against the DPPH radical. *Journal of Agricultural and Food Chemistry, 57*(3), 863–870. https:// doi.org/10.1021/jf803148z

Lucas, R., Comelles, F., Alcántara, D., Maldonado, O. S., Curcuroze, M., Parra, J. L., et al. (2010). Surface-active properties of lipophilic antioxidants tyrosol and hydroxytyrosol fatty acid esters: A potential explanation for the nonlinear hypothesis of the antioxidant activity in oil-in-water emulsions. *Journal of Agricultural and Food Chemistry, 58*(13), 8021–8026. https://doi. org/10.1021/jf1009928

Luo, S. Z., Chen, S. S., Pan, L. H., Qin, X. S., Zheng, Z., Zhao, Y. Y., et al. (2017). Antioxidative capacity of crude camellia seed oil: Impact of lipophilization products of blueberry antho-cyanin. *International Journal of Food Properties, 20*(2), 1627–1636. https://doi.org/10.108 0/10942912.2017.1350974

Mainini, F., Contini, A., Nava, D., Corsetto, P. A., Rizzo, A. M., Agradi, E., et al. (2013). Synthesis, molecular characterization and preliminary antioxidant activity evaluation of quercetin fatty esters. *JAOCS, Journal of the American Oil Chemists' Society, 90*(11), 1751–1759. https://doi. org/10.1007/s11746-013-2314-0

Moran, M. C., Pinazo, A., Perez, L., Clapes, P., Angelet, M., Garcia, M. T., et al. (2004). "Green" amino acid-based surfactants. *Green Chemistry, 6*(5), 233–240.

Nakanishi, K., Oltz, E. M., & Grunberger, D. (1989, February 9). Caffeic acid esters and methods of preparing and using same. Patent WO 89/00851, 33 pp.

Nakayama, T. (1995). Protective effects of caffeic acid esters against H_2O_2-induced cell damages. ASIC, 16th Colloque, Kyoto, Japan; ASIC: Paris; pp 372–379.

Nenadis, N., Zhang, H.-Y., & Tsimidou, M. Z. (2003). Structure-antioxidant activity relation-ship of ferulic acid derivatives: Effect of carbon side chain characteristic groups. *Journal of Agricultural and Food Chemistry, 51*, 1874–1879.

Neudörffer, A., Bonnefont-Rousselot, D., Legrand, A., Fleury, M.-B., & Largeron, M. (2004). 4-Hydroxycinnamic ethyl ester derivatives and related dehydrodimers: Relationship between oxidation potential and protective effects against oxidation of lowdensity lipoproteins. *Journal of Agricultural and Food Chemistry, 52*, 2084–2091.

Nihei, K.-I., Nihei, A., & Kubo, I. (2004). Molecular design of multifunctional food additives: Antioxidative antifungal agents. *Journal of Agricultural and Food Chemistry, 52*, 5011–5020.

Nkiliza, J. (1998, September 15). Process for esterifying a polyphenolic oligomeric extract of plant origin, composition thus obtained and use thereof. U.S. Patent 5,808,119, 3 pp.

Oh, W. Y., & Shahidi, F. (2017). Lipophilization of resveratrol and effects on antioxidant activities. *Journal of Agricultural and Food Chemistry, 65*, 8617–8625.

Oh, W. Y., & Shahidi, F. (2018). Antioxidant activity of resveratrol ester derivatives in food and biological model systems. *Food Chemistry, 261*(March), 267–273. https://doi.org/10.1016/j.foodchem.2018.03.085

Ortega-Valencia, J., García-Barradas, O., Bonilla-Landa, I., Mendoza-López, M., Luna-Solano, G., & Jimenez-Fernandez, M. (2019). Effect of lipophilization on the antioxidant activity of carvacrol, quercetin and vanillin with conjugated linoleic acid. *Revista Mexicana De Ingeniería Química, 18*(2), 637–645. https://doi.org/10.24275/uam/izt/dcbi/revmexingquim/2019v18n2/Ortega

Pande, G., & Akoh, C. C. (2016). Enzymatic synthesis of tyrosol-based phenolipids: Characterization and effect of alkyl chain unsaturation on the antioxidant activities in bulk oil and oil-in-water emulsion. *JAOCS, Journal of the American Oil Chemists' Society, 93*(3), 329–337. https://doi.org/10.1007/s11746-015-2775-4

Panya, A., Laguerre, M., Bayrasy, C., Lecomte, J., Villeneuve, P., McClements, D. J., et al. (2012). An investigation of the versatile antioxidant mechanisms of action of rosmarinate alkyl esters in oil-in-water emulsions. *Journal of Agricultural and Food Chemistry, 60*(10), 2692–2700.

Paximada, P., Echegoyen, Y., Koutinas, A. A., Mandala, I. G., & Lagaron, J. M. (2017). Encapsulation of hydrophilic and lipophilized catechin into nanoparticles through emulsion electrospraying. *Food Hydrocolloids, 64*, 123–132. https://doi.org/10.1016/j.foodhyd.2016.11.003

Pereira, D. M., Silva, T. C., Losada-Barreiro, S., Valentão, P., Paiva-Martins, F., & Andrade, P. B. (2017). Toxicity of phenolipids: Protocatechuic acid alkyl esters trigger disruption of mitochondrial membrane potential and caspase activation in macrophages. *Chemistry and Physics of Lipids, 206*, 16–27. https://doi.org/10.1016/j.chemphyslip.2017.05.013

Plat, T., & Linhardt, R. (2001). Syntheses and applications of sucrose-based esters. *Journal of Surfactants and Detergents, 4*(4), 415–421.

Priya, K., Venugopal, T., & Chadha, A. (2002). Pseudomonas cepacia lipase – Mediated trans-esterification reactions of hydrocinnamates. *Indian Journal of Biochemistry & Biophysics, 39*, 259–263.

Ramadan, M. F. (2008). Quercetin increases antioxidant activity of soy lecithin in a triolein model system. *LWT- Food Science and Technology, 41*(4), 581e587. https://doi.org/10.1016/j.lwt.2007.05.008

Ramadan, M. F. (2012). Antioxidant characteristics of phenolipids (quercetin-enriched lecithin) in lipid matrices. *Industrial Crops and Products, 36*(1), 363e369. https://doi.org/10.1016/j.indcrop.2011.10.008

Reddy, K. K., Ravinder, T., Prasad, R. B. N., & Kanjilal, S. (2011). Evaluation of the antioxidant activity of capsiate analogues in polar, nonpolar, and micellar media. *Journal of Agricultural and Food Chemistry, 59*(2), 564–569. https://doi.org/10.1021/jf104244m

Sasaki, K., Alamed, J., Weiss, J., Villeneuve, P., Lo'pez Giraldo, L. J., Lecomte, J., et al. (2010). Relationship between the physical properties of chlorogenic acid esters and their ability to inhibit lipid oxidation in oil-in-water emulsions. *Food Chemistry, 118*(3), 830–835.

Scholz, E., Heinrich, M., & Hunkler, D. (1994). Caffeoylquinic acids and some biological activities of Pluchea symphytifolia. *Planta Medica, 60*, 360–364.

Sellami, M., Châari, A., Aissa, I., Bouaziz, M., Gargouri, Y., & Miled, N. (2013). Newly synthesized dopamine ester derivatives and assessment of their antioxidant, antimicrobial and hemolytic activities. *Process Biochemistry, 48*(10), 1481–1487. https://doi.org/10.1016/j.procbio.2013.07.022

Shahidi, F., & Zhong, Y. (2011). Revisiting the polar paradox theory: A critical overview. *Journal of Agricultural and Food Chemistry, 59*, 3499–3504.

Silva, F. A. M., Borges, F., & Ferreira, M. A. (2001). Effects of phenolic propyl esters on the oxidative stability of refined sunflower oil. *Journal of Agricultural and Food Chemistry, 49*, 3936–3941.

Silva, F. A. M., Borges, F., Guimaraẽs, C., Lima, J. L. F. C., Matos, C., & Reis, S. (2000). Phenolic acids and derivatives: Studies on the relationship among structure, radical scavenging activity, and physicochemical parameters. *Journal of Agricultural and Food Chemistry, 48*, 2122–2126.

Sørensen, A.-D. M., Haahr, A.-M., Becker, E. M., Skibsted, L. H., Bergenståhl, B., Nilsson, L., et al. (2008). Interactions between iron, phenolic compounds, emulsifiers, and pH in omega-3-enriched oil-in-water emulsions. *Journal of Agricultural and Food Chemistry, 56*(5), 1740–1750.

Sørensen, A. D. M., Lyneborg, K. S., Villeneuve, P., & Jacobsen, C. (2015). Alkyl chain length impacts the antioxidative effect of lipophilized ferulic acid in fish oil enriched milk. *Journal of Functional Foods, 18*, 959–967. https://doi.org/10.1016/j.jff.2014.04.008

Sørensen, A. D. M., Nielsen, N. S., Yang, Z., Xu, X., & Jacobsen, C. (2012). Lipophilization of dihydrocaffeic acid affects its antioxidative properties in fish-oil-enriched emulsions. *European Journal of Lipid Science and Technology, 114*(2), 134–145. https://doi.org/10.1002/ejlt.201100002

Sørensen, A. D. M., Petersen, L. K., de Diego, S., Nielsen, N. S., Lue, B. M., Yang, Z., et al. (2012). The antioxidative effect of lipophilized rutin and dihydrocaffeic acid in fish oil enriched milk. *European Journal of Lipid Science and Technology, 114*(4), 434–445. https://doi.org/10.1002/ejlt.201100354

Sørensen, A. D. M., Villeneuve, P., & Jacobsen, C. (2017). Alkyl caffeates as antioxidants in O/W emulsions: Impact of emulsifier type and endogenous tocopherols. *European Journal of Lipid Science and Technology, 119*(6), 1–14. https://doi.org/10.1002/ejlt.201600276

Stamatis, H., Sereti, V., & Kolisis, F. N. (1999). Studies on the enzymatic synthesis of lipophilic derivatives of natural antioxidants. *Journal of the American Oil Chemists' Society, 76*, 1505–1510.

Stamatis, H., Sereti, V., & Kolisis, F. N. (2001). Enzymatic synthesis of hydrophilic and hydrophobic derivatives of natural phenolic acids in organic media. *Journal of Molecular Catalysis B: Enzymatic, 11*, 323–328.

Stasiuk, M., & Kozubek, A. (2010). Biological activity of phenolic lipids. *Cellular and Molecular Life Sciences, 67*(6), 841–860. https://doi.org/10.1007/s00018-009-0193-1

Szydłowska-Czerniak, A., Rabiej, D., & Krzemiński, M. (2018). Synthesis of novel octyl sinapate to enhance antioxidant capacity of rapeseed-linseed oil mixture. *Journal of the Science of Food and Agriculture, 98*(4), 1625–1631. https://doi.org/10.1002/jsfa.8637

Szydłowska-Czerniak, A., Rabiej, D., Kyselka, J., Dragoun, M., & Filip, V. (2018). Antioxidative effect of phenolic acids octyl esters on rapeseed oil stability. *LWT, 96*, 193–198. https://doi.org/10.1016/j.lwt.2018.05.033

Taniguchi, H., Nomura, E., Chikuno, T., & Minami, H. (1997, February 10). Antioxidant, cosmetic and new ferulic ester. Patent JP9040613.

Topakas, E., Stamatis, H., Biely, P., Kekos, D., Macris, B. J., & Christakopoulos, P. (2003). Purification and characterization of a feruloyl esterase from Fusarium oxysporum catalyzing esterification of phenolic acids in ternary water-organic solvent mixtures. *Journal of Biotechnology, 102*, 33–44.

Vercauteren, J., Weber, J.-F., Bisson, J.-L., & Bignon, J. (1994). Polyphenol derivative compositions and preparation thereof. Patent FR2706478.

Villeneuve, P., Muderhwa, J. M., Graille, J., & Haas, M. J. (2000). Customizing lipases for biocatalysis: A survey of chemical, physical and molecular biological approaches. *Journal of Molecular Catalysis B: Enzymatic, 4*, 113–148.

von Rybinski, W., & Hill, K. (1998). Alkyl polyglycosides; Properties and applications of a new class of surfactants. *Angewandte Chemie, International Edition, 37*(10), 1328–1345.

Wang, J., & Shahidi, F. (2014). Antioxidant activity of monooleyl and dioleyl p-coumarates in vitro and biological model systems. *European Journal of Lipid Science and Technology, 116*(4), 370–379. https://doi.org/10.1002/ejlt.201300348

Wang, M., Zhang, X., Zhong, Y. J., Perera, N., & Shahidi, F. (2016). Antiglycation activity of lipophilized epigallocatechin gallate (EGCG) derivatives. *Food Chemistry, 190*, 1022–1026. https://doi.org/10.1016/j.foodchem.2015.06.033

Wang, Y. Y., & Cao, X. J. (2011). Enzymatic synthesis of fatty acid ethyl esters by utilizing camellia oil soapstocks and diethyl carbonate. *Bioresource Technology, 102*, 10173–10179.

Wang, Z., Hwang, S. H., & Lim, S. S. (2016). Effect of novel synthesised policosanyl phenolates on lipid oxidation. *Czech Journal of Food Sciences, 34*(5), 414–421. https://doi.org/10.1722 1/530/2015-CJFS

Yarra, M., Kaki, S. S., Prasad, R. B. N., Mallampalli, K. S. L., Yedla, P., & Chityala, G. K. (2016). Synthesis of novel (Z)-methyl-12-aminooctadec-9-enoate-based phenolipids as potential antioxidants and chemotherapeutic agents. *European Journal of Lipid Science and Technology, 118*(4), 622–630. https://doi.org/10.1002/ejlt.201500160

Yu, X., Li, Y., & Wu, D. (2004). Enzymatic synthesis of gallic acid esters using microencapsulated tannase: Effect of organic solvents and enzyme specificity. *Journal of Molecular Catalysis B: Enzymatic, 30*, 69–73.

Yuji, H., Weiss, J., Villeneuve, P., Lopez Giraldo, L. J., FigueroaEspinoza, M. C., & Decker, E. A. (2007). Ability of surface-active antioxidants to inhibit lipid oxidation in oil-in-water emulsion. *Journal of Agricultural and Food Chemistry, 55*(26), 11052–11056.

Zheng, Y., Branford-White, C., Wu, X. M., Wu, C. Y., Xie, J. G., Quan, J., et al. (2010). Enzymatic synthesis of novel feruloylated lipids and their evaluation as antioxidants. *JAOCS, Journal of the American Oil Chemists' Society, 87*(3), 305–311. https://doi.org/10.1007/s11746-009-1496-y

Zhong, Y., Ma, C.-M., & Shahidi, F. (2012). Antioxidant and antiviral activities of lipophilic epigallocatechin gallate (EGCG) derivatives. *Journal of Functional Foods, 4*, 87–93.

Zhong, Y., & Shahidi, F. (2011). Lipophilized epigallocatechin gallate (EGCG) derivatives as novel antioxidants. *Journal of Agricultural and Food Chemistry, 59*, 6526–6533.

Index

Printed in the United States
by Baker & Taylor Publisher Services